SAVING SEEDS
The Economics of Conserving Crop Genetic Resources *Ex Situ* in the Future Harvest Centres of the CGIAR

Bonwoo Koo, Philip G. Pardey and Brian D. Wright

with

Paula Bramel, Daniel Debouck, M. Eric Van Dusen, Michael T. Jackson, N. Kameswara Rao, Bent Skovmand, Suketoshi Taba and Jan Valkoun

Prepared for the CGIAR System-wide Genetic Resources Programme

by the

International Food Policy Research Institute (IFPRI)

in collaboration with

Centro Internacional de Agricultura Tropical (CIAT)
Centro Internacional de Mejoramiento de Maíz y Trigo (CIMMYT)
International Center for Agricultural Research in the Dry Areas (ICARDA)
International Crops Research Institute for the Semi-Arid Tropics (ICRISAT)
International Rice Research Institute (IRRI)

CABI Publishing

CABI Publishing is a division of CAB International

CABI Publishing
CAB International
Wallingford
Oxfordshire OX10 8DE
UK

Tel: +44 (0)1491 832111
Fax: +44 (0)1491 833508
E-mail: cabi@cabi.org
Website: www.cabi-publishing.org

CABI Publishing
875 Massachusetts Avenue
7th Floor
Cambridge, MA 02139
USA

Tel: +1 617 395 4056
Fax: +1 617 354 6875
E-mail: cabi-nao@cabi.org

A catalogue record for this book is available from the British Library,
London, UK.

Library of Congress Cataloging-in-Publication Data
Koo, Bonwoo.
  Saving seeds : the economics of conserving crop genetic resources ex situ in
the future harvest centres of the CGIAR / Bonwoo Koo, Philip G. Pardey, and
Brian D. Wright ; with Paula Brame ... [et al.].
      p. cm.
  Includes bibliographical references and index.
  ISBN 0-85199-859-3 (alk. paper)
  1. Crops--Germplasm resources. 2. Gene banks, Plant. 3. Consultative Group
on International Agricultural Research. I. Pardey, Philip G. II. Wright, Brian,
1948 Jan. 1- III. Title.

  SB123.3.K66 2005
  631.5'23--dc22
                                                                    2004004194
ISBN 0 85199 859 3

Typeset by Columns Design Ltd, Reading
Printed and bound in the UK by Biddles Ltd, King's Lynn

SAVING SEEDS

The Economics of Conserving Crop Genetic
Resources *Ex Situ* in the Future Harvest Centres
of the CGIAR

The **Future Harvest** Centres comprise 15 food and environmental research organizations located around the world, which conduct research in partnership with farmers, scientists and policy makers to help alleviate poverty and increase food security while protecting the natural resource base. The Centres are supported by the Consultative Group on International Agricultural Research (CGIAR), a strategic alliance of countries, international and regional organizations, and private foundations that work with national agricultural research systems, the private sector and civil society. The CGIAR is co-sponsored by the International Fund for Agricultural Development (IFAD), the Food and Agriculture Organization of the United Nations (FAO), the United Nations Development Programme (UNDP), and the World Bank.

The **System-wide Genetic Resources Programme** (SGRP) joins the genetic resources programmes and activities of the Future Harvest Centres in a partnership whose goal is to maximize collaboration, particularly in five thematic areas. The thematic areas – policy, public awareness and representation, information, knowledge and technology, and capacity-building – relate to issues or fields of work that are critical to the success of genetic resources efforts. The SGRP contributes to the global effort to conserve agricultural, forestry and aquatic genetic resources and promotes their use in ways that are consistent with the Convention on Biological Diversity. The Inter-Centre Working Group on Genetic Resources (ICWG-GR), which includes representatives from the Centres and the Food and Agriculture Organization of the United Nations, is the Steering Committee. The International Plant Genetic Resources Institute (IPGRI) is the Convening Centre for SGRP. IPGRI is a Future Harvest Centre whose mandate is to advance the conservation and use of genetic diversity for the well-being of present and future generations. IPGRI's headquarters is in Maccarese, near Rome, Italy, with offices in another 22 countries worldwide.

The **International Food Policy Research Institute** (IFPRI) was established in 1975. IFPRI's mission is: to identify and analyse alternative national and international strategies and policies for meeting food needs of the developing world on a sustainable basis, with particular emphasis on low-income countries, poor people and sound management of the natural resource base that supports agriculture; to make the results of its research available to all those in a position to use them; and to help strengthen institutions conducting research and applying research results in developing countries. IFPRI is one of 15 Future Harvest agricultural research centres and receives its principal funding from governments, private foundations and international and regional organizations, most of whom are members of the Consultative Group on International Agricultural Research.

Contents

Contributors

Paula J. Bramel is currently involved as an independent consultant for a number of NGOs in the design and implementation of agricultural rehabilitation programmes. Prior to that she spent 11 years on the faculty of Kansas State University as a sorghum geneticist and 6 years at the International Crops Research Institute for the Semi-Arid Tropics (ICRISAT) as a genetic resources specialist and the head of the Genetic Resources programme. She has contributed to 57 refereed journal articles, five books and 27 book chapters. The ICRISAT Genebank was a main focus of her work, especially its efficient operation and the enhanced use of the conserved germplasm. The main focus of her research activities has been diversity assessment within the ICRISAT germplasm collections, assessing the value of the wild relatives of sorghum, pearl millet, chickpea, pigeonpea and groundnut, quantifying the degree and distribution of genetic diversity in the collection and the field, the use of molecular marker tools to describe and use genetic diversity, and developing procedures to better document environmental and farmers' knowledge for germplasm collections.
    Affiliation at the time of this study: ICRISAT.

Daniel Debouck was born in Brussels, Belgium, and holds a degree in engineering, a certificate in tropical agronomy, and a PhD in plant sciences from the Agronomical University of Gembloux, Belgium. Daniel is head of the Genetic Resources Unit at the Centro Internacional de Agricultura Tropical, where he works on the genetic diversity of New World crops. He has worked in crop physiology for the Ministry of Agriculture in Belgium and in genetic resources of grain legumes for the Food and Agriculture Organization of the United Nations.
    Affiliation at the time of this study: CIAT.

Michael T. Jackson is Director for Program Planning and Coordination at the International Rice Research Institute (IRRI) in the Philippines. A British national, he has spent more than 30 years working on the conservation and use of the genetic resources of potatoes, food and forage legumes, and rice. He has worked at the Centro Internacional de la Papa (CIP) in Peru and Central America, was a faculty member at the University of Birmingham, UK, and for 10 years was head of IRRI's Genetic Resources Center, home of the International Rice Genebank, until taking up his current position in 2001. He has a BSc in environmental botany and geography from the University of Southampton and MSc and PhD degrees in genetic resources and biosystematics from the University of Birmingham.

Affiliation at the time of this study: IRRI.

Bonwoo Koo is research fellow at the International Food Policy Research Institute (IFPRI). Before joining IFPRI, he was assistant research economist in the Agricultural Issues Center of the University of California. He received his MA and PhD in Agricultural and Resource Economics from the University of California at Berkeley in 1998. His main research area includes the economics of R&D and intellectual property rights, and the economics of biotechnology and genetic resources management.

Affiliation at the time of this study: IFPRI.

Philip G. Pardey, an Australian national, is Professor of Science and Technology Policy in the Department of Applied Economics, University of Minnesota. Prior to that he worked for 20 years in the CGIAR system, most recently at the International Food Policy Research Institute. Philip's research interests include agricultural productivity and R&D investments and impacts globally, the economics of science and technology generally, and a range of policy concerns regarding genetic resources and agricultural biotechnologies in particular. He currently serves on the US National Research Council's Standing Committee on Agricultural Biotechnology, Health and the Environment.

Affiliation at the time of this study: IFPRI at inception, University of Minnesota at conclusion.

N. Kameswara Rao, formerly the genebank curator at the International Crops Research Institute for the Semi-Arid Tropics (ICRISAT), is Germplasm Conservation Scientist at the International Plant Genetic Resources Institute. He has an MSc in botany from Andhra University, India, and a PhD in seed physiology from the University of Reading, UK. His main areas of interest are genetic resources conservation, especially on factors affecting seed longevity and genetic integrity. He also develops germplasm documentation and information management systems.

Affiliation at the time of this study: ICRISAT.

Bent Skovmand is currently Director of the Nordic Gene Bank in Sweden. Previously, he headed the Wheat Genetic Resources Program and coordinated all genetic resources activities at the Centro Internacional de Mejoramiento de Maíz y Trigo (CIMMYT) in Mexico. He was born in Denmark and earned his BSc, MSc and PhD degrees at the University of Minnesota in the United States. His research at CIMMYT focused on wheat and triticale, directed towards the conservation and utilization of *Triticum* genetic resources. Realizing the vulnerability of wheat and its relatives to the loss of agrobiodiversity, he saw the opportunity for CIMMYT to take the leading global role in *ex situ* conservation of genetic resources, and as a result CIMMYT now has a large and comprehensive collection of wheat genetic resources. In April 2003 he was awarded the Knight's Cross of the Order of Dannebrog by Queen Margrethe II of Denmark and in November of the same year he received the Frank Meyer Medal for Plant Genetic Resources, both for his lifelong achievements in wheat research and the conservation of wheat genetic resources.

Affiliation at the time of this study: CIMMYT.

Suketoshi Taba is Head of Maize Genetic Resources at the Centro Internacional de Mejoramiento de Maíz y Trigo (CIMMYT) in Mexico. Born in Okinawa, Japan, Suketoshi completed his graduate study in plant breeding at Kansas State University. He joined CIMMYT as a post-doctoral fellow in 1975, continuing to work with CIMMYT's maize breeding programme until 1986 when he undertook his current position. Suketoshi has developed the maize germplasm information management system used at CIMMYT and has regenerated and introduced more than 10,000 landrace accessions by collaborating with Latin American maize germplasm banks. He also now heads the prebreeding work of the maize programme as part of the genetic resources activity.

Affiliation at the time of this study: CIMMYT.

Jan Valkoun is Head of the Genetic Resources Unit at the International Center for Agricultural Research in the Dry Areas (ICARDA), located in Aleppo, Syria. He was born in Czechoslovakia and holds an MSc in Crop Production and Agronomy and a PhD in Plant Breeding and Genetics from the University of Agriculture, Prague. Jan's research focuses on the evaluation of the genetic diversity in natural populations of wild cereal progenitors and in existing *ex situ* collections using agromorphological and molecular characterization data; *in situ* conservation of wild relatives of wheat, barley and food legumes and wild forage legume species; and wheat prebreeding using wild progenitors and relatives.

Affiliation at the time of this study: ICARDA.

M. Eric Van Dusen is an agricultural and resource economist working on applied issues in the conservation and exchange of crop genetic resources. In 2000, he received a PhD from UC Davis with a dissertation on the *in situ* conservation of a traditional cropping system in an area of rural Mexico. In 2001, he worked as an agricultural economist at the United States Department of Agriculture, and as a consultant for FAO. Dr Van Dusen is currently a post-doctoral researcher at the University of California at Berkeley working on advancing the methodology for studies of on-farm conservation, on the impacts of international treaties on the exchange of crop genetic resources and on the international scope of intellectual property rights.

Affiliation at the time of this study: University of California, Davis.

Brian Wright's interest in agricultural economics dates from early experiences on his family's sheep station in the Riverina district of New South

Wales, Australia. He received his MA and PhD in Economics at Harvard University, taught at Yale University and is now Professor of Agricultural and Resource Economics at the University of California, Berkeley, where he is also Co-Director of Environmental Science. Brian's research interests include the economics of markets for storable commodities, agricultural policy, the economic dynamic of research and the economics of conservation and innovation of genetic resources.

Affiliation at the time of this study: University of California, Berkeley.

List of Tables

List of Figures

Foreword

The 20th century witnessed unprecedented and accelerating advances in crop breeding techniques, and widespread adoption of improved crop varieties by farmers the world over. These developments have had two key consequences. The rapid increase in crop yields in many parts of the world resulting from the use of improved crop varieties has generated billions of dollars of benefits to crop producers and consumers, and has helped to reduce poverty and malnutrition in many developing countries. However, the pervasive use of scientifically bred varieties is displacing the use of farmer-developed varieties, giving rise to concerns about 'genetic erosion' and a loss of agricultural biodiversity.

An important institutional initiative to address these concerns was the substantial expansion in the number of *ex situ* conservation facilities in the past few decades and the amount of crop germplasm stored within them. In 1998, it was estimated that over 6 million accessions (samples of seeds or other breeding materials, collectively denoted 'germplasm') were being conserved in more than 1300 genebanks worldwide. The 11 genebanks maintained by the Future Harvest Centres of the Consultative Group on International Agricultural Research (CGIAR) have become pivotal to the global conservation effort. They currently hold over 660,000 accessions of crops (plant or seed samples) grown mainly by poor farmers (such as cassava, millet, sorghum and cowpea), staple food crops consumed worldwide (including rice, wheat and maize) and tree species used in agroforestry systems.

Beyond the initial enthusiasm to build and stock these genebanks is the task of conserving these seeds and other germplasm for use by the current and far distant future generations. However, there are real concerns about the sustainability of these conservation efforts due to the mismatch between the generally short-term nature of the financial support for crop conservation and the long-term nature and intent of the undertaking. To sustain the conservation and use of agricultural germplasm held in *ex situ* genebanks, one significant option is to establish a financial endowment, the annual earnings of which could underwrite conservation efforts for the long haul.

In this volume, Bonwoo Koo, Philip Pardey, Brian Wright and collaborators at the Future Harvest Centres have undertaken a landmark assessment of the magnitude of these long-run conservation costs. They provide a detailed crop- and institution-specific accounting of those costs for the lion's share of the germplasm conserved within the CG system. They also investigate the sensitivity of these costs to changes in key parameters such as interest rates and the length of storage cycles, as well as other elements such as the size, scope and location of the holdings.

This book (including its comprehensive data appendices) provides an invaluable guide for those seeking to conduct similar analyses for various management and policy purposes. More immediately, these costing results are being used to benchmark current efforts to establish an endowment fund, known as the Global Crop Diversity Trust, designed to underwrite internationally significant crop conservation efforts in perpetuity. As the authors of this report conclude, juxtaposing the sizeable but comparatively modest costs of germplasm conservation against the billions of dollars realized from the use of improved crop varieties furnishes 'an obvious and compelling case for aligning the very long-run nature of these conservation efforts with an endowment or equivalent funding mechanism'. Such an investment will help to assure future generations of farmers and crop breeders access to the crop samples currently conserved – an invaluable legacy of crop genetic diversity developed by scientists and the generations of farmers that have gone before us.

Joachim von Braun
Director General, IFPRI
Emile Frison
Director General, IPGRI and Programme Leader, SGRP
Geoffrey Hawtin
Interim Executive Secretary, Global Crop Diversity Trust

Acknowledgements

In conducting this study we received lots of excellent help from lots of people. We are especially grateful for the assistance we obtained from Anne Acosta, Arnoldo Amaya, Claudio Cafati, Jaime Diaz, Jesse Dubin, Paul Fox, Lucy Gilchrist, Arne Hede, Rafael Herrera, Alejandro López, Francisco Magallanes and Peter Ninnes from CIMMYT; Bilal Humeid, Suresh Sitaraman, Jurgen Diekmann, Jan Konopka, Siham Asaad, Ali Ismail, Ali Shehaden and F. Sweid from ICARDA; R. Ortiz, B. Kumar, D.V.S.S.R. Sastry, V. Reddy, K. Reddy and S. Singh from ICRISAT; Gordon MacNeil, Flora de Guzman, Renato Reano, Socorro Almazan, Amita Juliano, Elizabeth Naredo and Adel Alcantara from IRRI; Alba Marina, Graciela Mafla, Roosevelt Escobar, Benjamin Pineda and Carmenza Llano from CIAT in gathering and helping us understand the nuances in the detailed data assembled for the five CG centres directly surveyed in the course of our research.

We are also grateful for the assistance Tony Simons provided concerning conservation costs at ICRAF and to Samy Gaiji for help with the SINGER (System-wide Information Network for Genetic Resources) data. Martin Van Weerdenberg of IFPRI and Vince Smith of Montana State University helped us develop the annualized estimates of capital costs used in this study. Jane Toll and Geoffrey Hawtin of IPGRI/SGRP, Michelle Marra of North Carolina State University and Melinda Smale of IFPRI/IPGRI provided particularly helpful advice during the course of the project and feedback on draft sections of this report. We received much practical assistance from the CGIAR System-wide Genetic Resources Programme (SGRP) staff, especially Layla Daoud. We would also like to acknowledge the main financial support provided by the governments of Japan, The Netherlands and Switzerland, along with the World Bank through the SGRP, with in-kind contributions from CGIAR centres and additional support from the University of Minnesota and the Giannini Foundation at the University of California, Berkeley. Finally, the considerable editorial assistance we received from Mary-Jane Banks was exceptionally helpful, and for that we hereby record our debt of gratitude.

Acronyms and Abbreviations

AFLP	amplified fragment length polymorphism
AICSIP	All India Coordinated Sorghum Improvement Project
AWCC	Australian Winter Cereals Collection
CAN/PGR	Central Asia Network for Plant Genetic Resources
CATN/PGR	Central Asia and Transcaucasian Network for Plant Genetic Resources
CGIAR	Consultative Group on International Agricultural Research
CIAT	Centro Internacional de Agricultura Tropical
CIFOR	Center for International Forestry Research
CIMMYT	Centro Internacional de Mejoramiento de Maíz y Trigo
CIP	Centro Internacional de la Papa
CWANA	Central and West Asia and North Africa
ELISA	enzyme-linked immunosorbent assay
FAO	Food and Agriculture Organization of the United Nations
FESP	Farm and Engineering Service Program
FIA	Federal Institute of Agrobiology
GEF	Global Environmental Facility
GHL	Germplasm Health Laboratory
GRIP	Genetic Resources Information Package
GRU	Genetic Resources Unit
IARI	Indian Agricultural Research Institute
IBPGR	International Board for Plant Genetic Resources
ICAR	Indian Council of Agricultural Research
ICARDA	International Center for Agricultural Research in the Dry Areas
ICLARM	WorldFish Center (formerly International Center for Living Aquatic Resources Management)
ICRAF	World Agroforestry Centre (formerly International Centre for Research in Agroforestry)

ICRISAT	International Crops Research Institute for the Semi-Arid Tropics
IFPRI	International Food Policy Research Institute
IITA	International Institute of Tropical Agriculture
ILRI	International Livestock Research Institute
INIA	Instituto Nacional de Investigaciones Agrícolas
INIBAP	International Network for the Improvement of Banana and Plantain
IPGRI	International Plant Genetic Resources Institute
IRD	Institut de Recherche pour le Développement
IRG	International Rice Genebank
IRGCIS	International Rice Genebank Collection Information System
IRRI	International Rice Research Institute
ISNAR	International Service for National Agricultural Research
IWIS	International Wheat Information System
IWMI	International Water Management Institute
LAMP	Latin American Maize Project
MZBANK	Maize Germplasm Bank Management System
NARS	national agricultural research systems
NAS	National Academy of Sciences
NBPGR	National Bureau of Plant Genetic Resources
NIAR	National Institute of Agrobiological Resources
NRC	National Research Council
NSSL	National Seed Storage Laboratory
OSS	Office of Special Studies
PCR	polymerase chain reaction
PQU	Plant Quarantine Unit
RAC	Station Fédérale de Recherches Agronomiques de Changins
RAPD	randomly amplified polymorphic DNA
SDC	Swiss Agency for Development and Cooperation
SHU	Seed Health Unit
SINGER	System-wide Information Network for Genetic Resources
WARDA	Africa Rice Center (formerly West Africa Rice Development Association)

Dedication

This monograph is dedicated to the farmers and genebank managers who painstakingly nurture the crop diversity that underpins the world food supply. We, like those of future generations worldwide, owe them a debt of gratitude.

1 Introduction

BONWOO KOO, PHILIP G. PARDEY AND BRIAN D. WRIGHT

Genebanks are a very recent institutional innovation to conserve germplasm, the 'material that controls heredity' (Witt, 1985, p. 8). For most of agriculture's 10,000-year history, it was farmers who saved seeds from one season for planting in the next. The idea of setting aside seeds from around the world in special facilities for global use by breeders and others in the near and distant future did not really take hold until the early 20th century. The credit for this idea and its implementation largely rests with the famous Russian biologist Nikolai Vavilov. During three decades of travel over five continents, he amassed the largest collection of species and strains of cultivated plants in the world (at that time) and developed theories on how to use this material for breeding improved varieties (Reznik and Vavilov, 1997). Seed collections held by breeders and researchers have been expanded and organized into more comprehensive *ex situ* genebanks (meaning storage facilities 'away from the source') that focus on particular classes of crops.

Vavilov's principal concern was crop improvement, rather than conservation. Investment in long-term conservation is an even more recent phenomenon. Pistorius (1997) identifies the US National Seed Storage Laboratory (NSSL) at Fort Collins, Colorado, created in 1958, as the first such facility. Since then, a sizeable investment has been made in collecting and conserving landraces (farmer-developed varieties) and wild and weedy species of crops in genebanks around the world. Motivating these investments were concerns that the genetic basis of agriculture – whether for commercial or subsistence production – was narrowing globally for many agricultural crops with the advent of more genetically uniform but superior-performing varieties developed at an accelerating pace beginning in the 1960s.[1]

Recent estimates quantified existing global *ex situ* collections at over 6 million accessions in more than 1300 genebanks worldwide (FAO, 1998). About 10% of these accessions are maintained within the centres of the Consultative Group on International Agricultural Research (CGIAR), most of them as 'in trust' accessions for the international community under the auspices of the Food and Agriculture Organization of the United Nations (FAO).[2] Since the 1970s, the 11 genebanks maintained by the CGIAR (or CG for short) have become pivotal to the global conservation effort, currently holding over 660,000 accessions of crops (plant or seed samples) grown mainly by poor farmers (like cassava, millet, sorghum and cowpea), staple food crops consumed worldwide (like rice, wheat and maize) and tree species used in agroforestry systems (Table 1.1).

The number of modern, *ex situ* conservation facilities has grown over the past several decades, and the technology for storing germplasm has dramatically improved; but with the focus on performance enhancement and capacity expansion, key management questions have been overlooked. These include what and how much should be conserved; where should it be stored and regenerated when required; and how is conserved germplasm used and how should it be used?

These questions all have economic dimensions, although answering them with any precision is problematic.[3] First, estimating the marginal benefits of conserving each type of genebank accession is an important, but particularly difficult, element, in part because attributing an appropriate part of the agronomic improvement in a plant to the use of conserved germplasm is a daunting, if not intractable, inferential challenge (see, for example, Pardey *et al.*, 1996, 2002). Secondly, many modern genebank facilities are so new that insufficient time has elapsed for breeders to establish a usable time series of realized gains attributable to their establishment. Beyond immediate agronomic values that are estimable in principle, germplasm also has value in terms of as yet unidentified future demand ('option value') and the sheer value of its very existence as opposed to extinction ('existence value').[4] Though methodologies do exist to assess the overall economic benefits from conserving seed, empirical results are bound to be imprecise.

The cost side, on the other hand, predominantly involves items that are at least estimable, in principle, from historical data relevant to existing genebank operations. If the total and marginal costs of the genebank operations are judged to be less than any reasonable lower-bound estimate of the corresponding benefits, then it may not be necessary to confront the challenge of precisely estimating the latter to establish the economic justification of the genebank operation. These rationales have motivated a series of detailed costing studies over the past several years, led by the International Food Policy Research Institute (IFPRI) in close collaboration with colleagues at five CGIAR genebanks.

The structure of conservation costs depends critically on: (i) the type of crops being conserved; (ii) institutional differences such as cost-sharing arrangements within each CG centre; and (iii) the local climate and general state of the infrastructure available to each genebank (such as electricity

Table 1.1. Size and structure of the germplasm collection at the CGIAR centres, 2001. (In-trust figures provided by IPGRI and totals provided directly by individual genebanks during 2001.)

Centre (location)	Crop	Number of accessions		
		In-trust	Other	Total
CIAT (Colombia)	Cassava	5,728	2,332	8,060
	Common bean	30,590	810	31,400
	Forages	16,339	7,845	24,184
	Total	**52,657**	**10,987**	**63,644**
CIMMYT (Mexico)	Wheat	79,912	75,000[a]	154,912
	Maize	20,411	4,675	25,086
	Total	**100,323**	**79,675**	**179,998**
CIP (Peru)	Potato	5,057	2,582	7,639
	Sweet potato	6,413	1,246	7,659
	Andean roots/tubers	1,112	383	1,495
	Total	**12,582**	**4,211**	**16,793**
ICARDA (Syria)	Cereals	54,218	5,795	60,013
	Forages	24,581	5,947	30,528
	Chickpea	9,116	2,103	11,219
	Lentil	7,827	2,135	9,962
	Faba bean	9,074	1,671	10,745
	Total	**104,816**	**17,651**	**122,467**
ICRAF (Kenya)	**Agroforestry trees**	**25**	**10,000[a]**	**10,025**
ICRISAT (India)	Sorghum	35,780	941	36,721
	Pearl millet	21,250	142	21,392
	Pigeonpea	12,698	846	13,544
	Chickpea	16,961	289	17,250
	Groundnut	14,357	985	15,342
	Minor millets	9,050	205	9,252
	Total	**110,096**	**3,405**	**113,501**
IITA (Nigeria)[b]	Bambara groundnut	2,029	–	2,029
	Banana	–	400	400
	Cassava	2,158	1,371	3,529
	Cowpea	15,001	1,000	16,001
	Soybean	1,909	1,144	3,053
	Wild *Vigna*	1,634	50	1,684
	Miscellaneous legumes[c]	–	400	400
	Yam	2,878	822	3,700
	Total	**25,609**	**5,187**	**30,796**
ILRI (Kenya)	**Forages**	**11,537**	**1,667**	**13,204**
IPGRI/INIBAP (Italy)	***Musa***	**914**	**229**	**1,143**
IRRI (Philippines)	Cultivated rice	77,827	16,737	94,564
	Wild rice	2,790	1,778	4,568
	Total	**80,617**	**18,515**	**99,132**
WARDA (Côte d'Ivoire)[d]	**Rice**	**14,917**	**460**	**15,377**
CG Total		**514,093**	**151,987**	**666,080**

[a]Estimate provided by the genebank manager.
[b]IITA holds about 2500 accessions of maize and multipurpose trees in addition to the material stipulated in this table.
[c]Includes African yam bean, Kersting's groundnut and various beans (e.g. lablab, jack and winged beans).
[d]The WARDA base collection is housed at IITA.

supplies, communications and international shipment options). For example, regenerating cross-pollinating crops (like maize, sorghum and pearl millet) or wild and weedy species is typically more complicated than regenerating self-pollinating cultivated species.[5] As demonstrated below, vegetatively propagated species maintained as clones *in vitro* and in field genebanks are much more expensive to conserve than stored seeds. Besides these crop-specific aspects, differences in wage structures and the composition of labour (which are affected by local labour laws and practices) also have significant impacts on the overall costs. Moreover, if the local climate is inappropriate for regenerating some varieties, it may be necessary to plant them out at other locations or expend resources on means of climate modification (such as greenhouse structures).

The implications of the above variations mean that different genebanks will have different cost structures. To assess the range of cost structure within a reasonable time frame, we selected five CG centres for in-depth studies, standardizing the treatment of the data as far as possible to facilitate meaningful comparisons. The five centres are Centro Internacional de Agricultura Tropical (CIAT), Centro Internacional de Mejoramiento de Maíz y Trigo (CIMMYT), International Center for Agricultural Research in the Dry Areas (ICARDA), International Crops Research Institute for the Semi-Arid Tropics (ICRISAT) and International Rice Research Institute (IRRI), constituting about 16 different species of crops that include more than 87% of the total collection held by the CGIAR (578,742 of 666,080 accessions, as shown in Table 1.1).[6]

In estimating the costs of different genebank activities for each crop, these studies provide the basis for solving important genebank management issues. These include choice of cost-effective storage methods, setting fees for distribution services and determining economies of scale.

Germplasm conservation is a long-run proposition, more accurately described as a commitment 'in perpetuity'. Concerns over declining financial resources for *ex situ* genebanks have instigated interest in an independent and permanent funding mechanism;[7] the growing mismatch between short-term financial support and the long-term intent of the conservation effort validate the seriousness of these concerns. Thus, a further purpose of this study is to estimate the necessary size of a conservation endowment fund to enable the current level of CG genebank operations to continue permanently. Endowing a fund to support the *ex situ* conservation of genetic resources is like investing in land set aside to preserve biodiversity *in situ* (meaning 'at the source'). Both require an upfront injection of investment capital to ensure the maintenance of biological diversity over the long haul.

The next chapter outlines the simple economics of genebank costing, along with the methodology and assumptions used for the series of costing studies presented and examined in Chapters 3–7.[8] Along with discussion of a few management issues of genebank operation, Chapter 8 compares the results from the five case studies and extrapolates the estimated size of the endowment fund needed to conserve and distribute the 11 CG genebanks' current levels of holdings in perpetuity.

Notes

[1]Concerns about 'genetic erosion' (loosely, a narrowing of the genetic resource base used by farmers or breeders for improving crop varieties) were raised by the outbreak of southern corn leaf blight in the USA in the 1970s, and were addressed by NRC (1972) and Harlan (1972), among others. However, the seriousness of this issue varies from crop to crop. NRC (1972) found common beans to be 'impressively uniform and impressively valuable', whereas Smale *et al.* (2002) conclude that 'the data are not consistent with the hypothesis that the genetic base of CIMMYT germplasm [of spring bread wheat] has tended to narrow over time'.

[2]The CGIAR 'in-trust' agreement was signed in October 1994, wherein the CG centres agreed to hold so-called designated germplasm in trust for the international community under the auspices of the FAO. Designated material is made freely available for research and crop improvement purposes, provided germplasm recipients abide by a material transfer agreement in which they eschew intellectual property rights over the germplasm shipped from CG centres. For a succinct statement of the in-trust agreement and related material transfer agreements, see Fowler (2003).

[3]Frankel *et al.* (1995) offer some technical (non-economic) perspectives on many of these same issues.

[4]Koo *et al.* (2003a) discuss option values in the context of improved seed varieties in China. See also Gollin *et al.* (2000), Koo and Wright (2000) and Zohrabian *et al.* (2003) for ideas on valuing and evaluating genetic resources used in crop improvement research.

[5]It is crucial to regenerate material in ways that minimize the genetic drift from the planted to the harvested sample. In promiscuously outcrossing plants like maize, for example, the plant producing more pollen would tend to fertilize more flowers and so increase its share of seed samples after regeneration. Fairly elaborate procedures, such as hand-pollinating each plant and isolating the pollen of each plant by placing a cover over its tassels, are needed to prevent this.

[6]For descriptions of developments regarding the CGIAR, see Baum (1986), Gryseels and Anderson (1991) and Anderson and Dalrymple (1999).

[7]Currently, the CG genebanks are financed from short-term (often year-by-year) pledges of support to the system and its centres by its members and from project funds with limited lifespans (sometimes 5 years, but often 3 years or less).

[8]Note that, based on the timing of the centre studies, the chapters are presented chronologically. For the CIMMYT study, data are from 1996; for ICARDA, data are from 1998; for IRRI and ICRISAT, data are from 1999; and for CIAT, data are from 2000.

2 The Economics of Genebank Costing*

BONWOO KOO, PHILIP G. PARDEY AND BRIAN D. WRIGHT

Saving seeds in a genebank is not at all like storing books in a library or maintaining a museum of history or antiquities; it is perhaps more akin to keeping animals in a zoo or maintaining a botanical garden. Seeds and other genetic material stored as plantlets in special growth mediums are living entities, requiring special scientific expertise and continuous care to ensure their health and long-term survival. Costing *ex situ* conservation efforts in sufficient detail to be of practical and policy use requires a fairly detailed understanding of the technical requirements, as well as access to correspondingly detailed financial records. It also requires an understanding of some basic economic principles to identify different cost components so that the implications of alternative scientific management and conservation options can be assessed. Judicious use of such cost data can reveal potential efficiency gains from doing things differently as well as providing the basis for estimating the funds required to conserve germplasm over the long haul.

Overview of the Genebank Operation

Storing seeds and other plant material

Most of the seed samples (or accessions) stored in genebank facilities are placed in packets or small containers and held in medium-term storage facilities (maintained between 0 and 5°C and 15 and 20% relative humidity) as an active collection. Much of this material is also kept in long-term storage

*This chapter is a modified version of Koo, B., Pardey, P.G. and Wright, B.D., The economic costs of conserving genetic resources at the CGIAR centres, *Agricultural Economics*, Vol. 29, pp. 287–297, Copyright 2003, with permission from Elsevier.

facilities as a base collection (at colder temperatures, often ranging from -18 to $-20°C$). The consensus is that most but perhaps not all seed samples – hence the need for vigilance – will remain viable for 20–30 years in medium-term storage and for up to 100 years in long-term storage, depending on the type of species, the initial seed quality and the specifics of the storage environment. Seed samples are checked regularly for viability, every 5–10 years, and regenerated if the viability drops below a threshold level.

Vegetatively propagated species (including crops like cassava, potatoes and bananas) are conserved as whole plants in field genebanks. They are also kept as live specimens, often maintained on a special growth medium in test tubes stored under warm, lighted conditions ($23°C$ and 1500–2000 lux) in so-called *in vitro* genebanks. Plants in field genebanks can be readily characterized and evaluated but are susceptible to environmental variations. They are also difficult to distribute internationally because of increasingly stringent phytosanitary restrictions. *In vitro* genebanks store plants in controlled environments, with less risk of natural disaster, and facilitate the distribution of disease-free materials internationally. Another option for long-term conservation that may become economically attractive is the use of cryoconservation techniques, whereby plant material (or sometimes even seeds) is conserved at extremely low temperatures ($-196°C$ maintained with liquid nitrogen). Some material is stored this way now; however, cryoconservation protocols for many species, and even some genotypes within a species, have yet to be fully determined, though they remain under active investigation.

The protocols used by the World Agroforestry Centre (ICRAF, formerly known as the International Centre for Research in Agroforestry) for conserving and distributing tree germplasm are quite different from the protocols generally used for crop species throughout the rest of the CGIAR. Some tree species are kept as seed in cold storage (much like other crops, with the exception that the amount of material stored per accession is often vastly larger than for other crops), but other material is conserved in field genebanks with the bulk of the distributions made from seed harvested from 'nuclear or catalyst stocks' maintained at various locations throughout the world.

Shipping seeds and other plant material

Complementing the conservation services, another important service provided by genebanks, especially CG genebanks, is the dissemination of seed and other plant samples free of charge upon request. Samples for ready dissemination are maintained in medium-term storage as active collections, which require more frequent viability testing and regeneration. Table 2.1 shows the total number of seed samples disseminated from individual CG genebanks for a recent 7-year period. From 1994 to 1999, nearly 600,000 samples were disseminated by the CG genebanks (an average of about 100,000 samples per year), of which more than half were sent to breeders and other scientists working within each centre.

Table 2.1. Number of samples disseminated from CG genebanks, 1994–2000. (The data for ILRI and IRRI in year 2000 are obtained from the SINGER database, and the remaining data from surveys of genebank managers.)

Centre	Crop	1994	1995	1996	1997	1998	1999	2000[a]
CIAT	Cassava	550	527	149	219	366	460	2,176
	Forages	3,231	1,133	1,320	1,053	518	525	517
	Common bean	8,877	7,565	8,705	10,481	8,493	9,600	4,256
	Total	**12,658**	**9,225**	**10,174**	**11,753**	**9,377**	**10,585**	**6,949**
CIMMYT	Maize	4,393	3,338	3,685	2,598	5,062	2,831	3,565
	Wheat	2,244	460	1,835	9,974	22,105	6,512	n/a
	Total	**6,637**	**3,798**	**5,520**	**12,572**	**27,167**	**9,343**	**3,565**
CIP	Potato	8,333	7,148	5,314	2,142	5,166	1,878	2,154
	Sweet potato	1,655	3,808	2,046	1,818	1,678	498	444
	Andean roots/tubers	242	–	28	–	–	–	66
	Total	**10,230**	**10,956**	**7,388**	**3,960**	**6,844**	**2,376**	**2,664**
ICARDA	Cereal	12,646	10,074	13,502	10,323	8,916	11,720	8,001
	Chickpea	5,248	5,575	5,437	7,066	5,111	2,812	2,090
	Lentil	5,464	3,849	3,994	3,978	3,911	3,286	3,057
	Faba bean	412	2,393	1,601	1,434	3,917	3,306	2,286
	Forages	8,883	7,983	9,246	8,777	7,696	9,178	5,193
	Total	**32,653**	**29,874**	**33,780**	**31,578**	**29,551**	**30,302**	**20,627**
ICRISAT	Chickpea	9,329	2,893	9,778	3,283	7,046	6,756	3,003
	Groundnut	6,180	3,737	3,443	4,787	2,475	5,604	4,872
	Pearl millet	2,301	3,143	2,695	1,224	1,344	1,980	2,671
	Minor millets	3,912	402	145	462	121	452	53
	Pigeonpea	6,520	2,206	2,866	1,014	962	1,595	2,657
	Sorghum	7,924	2,983	3,525	4,667	4,731	5,456	5,865
	Total	**36,166**	**15,364**	**22,452**	**15,437**	**16,679**	**21,843**	**19,121**
IITA	Bambara groundnut	118	50	147	19	29	16	14
	Cassava	347	330	690	1,871	1,211	461	493
	Cowpea	217	323	95	524	387	12,501	602
	Soybean	198	894	2	22	111	202	17
	Wild *Vigna*	272	393	71	18	286	93	157
	Miscellaneous legumes	318	148	136	94	96	34	38
	Yam	213	303	250	214	268	257	132
	Total	**1,683**	**2,441**	**1,391**	**2,762**	**2,388**	**13,564**	**1,453**

Continued

Table 2.1. *Continued.*

Centre	Crop	1994	1995	1996	1997	1998	1999	2000[a]
ILRI[b]	Forages	2,355	2,240	1,591	1,954	2,267	2,139	1,313
IPGRI/INIBAP	*Musa*	n/a	69	83	81	69	88	91
IRRI[b]	Rice	25,802	15,630	10,958	5,633	6,670	6,194	3,516
WARDA	Rice	304	718	622	2,437	312	123	126
CG Total		**128,488**	**90,315**	**93,959**	**88,167**	**101,324**	**96,557**	**59,425**

[a]Year 2000 data of wheat shipments from CIMMYT are missing, partly due to an incomplete accounting.
[b]All ILRI and year 2000 IRRI data represent samples shipped externally from each centre.
n/a, not available.

Most of the samples held in the CG genebanks are landraces and wild species. This material is an important source of genetic diversity (and a potentially valuable source of novel and useful traits), but it is currently less accessible for use in crop-breeding programmes because of the limited knowledge about specific traits that might exist in given varieties and the cost of expressing any such traits in advanced breeding lines or commercial varieties ready for release to farmers. Hence the demand from breeders for this type of material is lower than the demand for well-characterized and better-known breeding lines. While a substantial number of samples have been distributed, that number may be an inaccurate indication of actual usage. To reasonably assess the use value of the material held in the CG genebanks, more complete information is needed on its impact on global crop-breeding efforts as sources of new, desirable traits among other uses.

Genebank services and the scope of this study

For this costing analysis, we grouped the genebank operations into a set of three main service categories: conservation, distribution and information. Conservation services include conserving agricultural genetic diversity in the form of a 'base collection' held in controlled environmental conditions to maintain the stored plants (or plant parts) and seeds for use in the distant future. Properly fulfilling this function requires maintaining healthy (disease-free) and viable germplasm in long-term storage; periodically checking viability (via germination tests) and regenerating stored material when required (planting the aged seeds and storing their progeny); and maintaining duplicates of the collection at other locations for safety. Basic conservation activities also require keeping track of the size and condition of each holding and documenting 'passport' data that indicate the source of the sample (such as another genebank or institution or a field expedition) and its physical attributes (including plant height; seed characteristics like size, colour and shape; and evident pest and disease susceptibility). Much of this agronomic information is collected when the seeds are grown in greenhouses or in the field for disease screening or regeneration purposes.

The distribution activities relate to fulfilling requests for specific samples. This typically involves maintaining an 'active collection' of germplasm in a medium-term storage facility from which samples – of seed or *in vitro* plantlets of (usually vegetatively propagated) crops like cassava – are disseminated to researchers, crop breeders, farmers and other genebanks. Material stored in active collections typically requires more frequent regeneration, first, because the environment in active storage facilities is less conducive to long-term conservation (controlled temperatures and humidity, for example, are often not held as low or are less stable given frequent access to retrieve samples for distribution) and, secondly, because seed sample sizes are incrementally reduced and eventually need replenishment even if the seed is still viable.

Information services provide a range of useful and reliably accessible data about each accession to expedite the use of material for crop-improvement or other research purposes. Some of this information is obtained by purposefully screening the genebank collection for varieties with resistance to certain pests and diseases (often by planting out many varieties and exposing them to the pest or disease). Increasingly modern biotechnology tools are also used to collect data at the molecular level, identifying the genetic basis for certain traits and other genetic information deemed desirable in breeding programmes.

The demarcation between genebank and breeding functions is not always clear-cut. In some settings (like the CG centres where genebank activities form part of a more comprehensive research operation), some of the information services emanate from crop-breeding programmes. In other cases, some of the prebreeding activities that are typically done as part of a breeding programme fall under the realm of a genetic resource or genebank programme. To meaningfully compare among centres using a consistent set of core conservation activities, we confined the scope of our costing exercise to those functions that are essential for fulfilling the conservation and distribution demands placed on a genebank. Table 2.2 provides an overview of the functions that may form part of a genetic resource programme and identifies the subset of activities included in our costing exercise. Notably, some management aspects deal with genetic resource issues not directly included in the conservation and distribution activities costed; hence only a fraction of the total management costs was included in the calculations.

Table 2.2. A categorization of genebank operations.

Costed activities	Uncosted activities
Management	Management
● Administrative tasks	● Administrative tasks
● Data-related activities[a]	● Data-related activities[a]
Conservation	Information
● Acquisition (including basic morphological and passport data)	● Characterization (additional morphological and molecular characterization)
● Long-term storage	● Evaluation
● Safety duplication	● Prebreeding
● Viability testing	● Other research
● Regeneration	
Distribution	Other services
● Medium-term storage	● Germplasm collection
● Dissemination	● Training
● Viability testing	
● Regeneration	

Note: Some 'management' tasks deal with genetic resource issues not directly included in the conservation and distribution activities costed; hence only a share of the total management costs was included in the cost calculations.
[a]Excludes system-wide documentation and dissemination activities, such as from the SINGER database.

The Simple Economics of Genebanking

To structure the costing exercise we considered the genebank operations within a production economics framework, wherein inputs such as labour, buildings, equipment and acquired seeds are processed to produce outputs in the form of stored and distributed seeds and the information that accompanies them. Properly stored seeds and related information can be disseminated on demand for current use or held in storage as use options that can be exercised, repeatedly if necessary, in future years.

We also partitioned total costs into their variable, capital and quasi-fixed components, as outlined below, before summarizing each category in terms of average and marginal costs:

1. Variable inputs are sensitive to the size of the operation and encompass labour inputs, such as technicians and temporary workers, who are often paid on a daily basis, as well as operational inputs, such as power, and various material inputs.
2. Capital inputs are insensitive to the size of the operation and include buildings and durable equipment.
3. Quasi-fixed inputs are neither fixed nor variable but 'lumpy', meaning they are indivisible units that cannot be easily apportioned and attuned to marginal changes in the size of genebank operations but are more variable than capital items. In this framework, quasi-fixed inputs include the 'human capital' costs of skilled labour and scientific expertise, such as genebank managers and laboratory scientists.

This classification makes it possible to investigate the magnitude of possible economies of scale, which, loosely speaking, refer to reductions in the unit costs of output (in this case stored or distributed seeds) that come with increases in the size of the operation. The phenomenon reflects factors such as the decreasing relation of surface area to volume of the refrigerated facility and the appropriate division of labour. In larger operations, comparatively well-paid geneticists or agronomists can be fully employed in managing the genebank rather than spending time on less productive tasks – such as sorting and classifying seed – which are better assigned to less expensive technicians or temporary workers. Economies of scale are further exploited when the genebank facility is large enough to use efficiently other lumpy fixed factors – such as physical infrastructure and scientific expertise – as well as variable inputs, such as hired labour and chemicals.

Figure 2.1 relates the typical changes in average and marginal costs to changes in the amount of output (for example, the number of accessions stored during a given period). Average fixed (or quasi-fixed) costs generally decline as output increases – as when a given fixed cost, such as the cost of the genebank facility, is charged against a greater amount of output, such as more stored seeds. Marginal costs are the addition to total costs from the addition of the last unit of output – commonly marginal costs eventually increase under the law of diminishing marginal returns. In Fig. 2.1 the number of accessions identified on the *x*-axis could equally refer to the number of accessions stored, regenerated or disseminated in any particular year.

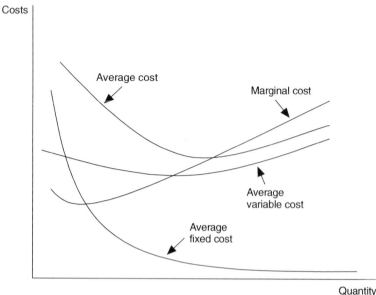

Fig. 2.1. Average and marginal cost curves for conserving seed.

When inputs are used efficiently, as the relative prices of inputs such as labour and chemicals change, so should the mix of those inputs in the conservation and distribution of seed. The sensitivity of the mix of inputs to changes in the relative prices of inputs is in turn dependent on the degree of substitutability or complementarity of the respective inputs. An increase in the price of labour over time, for instance, should spur a substitution of other inputs for labour – such as improved refrigeration equipment that lengthens storage life and in turn the regeneration cycle for stored seeds, ultimately reducing the labour requirement.

Changes in the technology of genebank operations also affect the optimal amount, mix and cost of inputs used in the longer run. If international seed distribution becomes quicker and cheaper through improvements in express mail, it could substitute for duplicate conservation facilities around the world (if other barriers are not unduly burdensome). Moreover, the direction and nature of the change in the technologies available may themselves be driven by shifts in relative prices – the so-called induced-innovation model of technical change (see Hayami and Ruttan, 1985). Over the longer run, technical changes will tend to reinforce the magnitude and direction of the shorter-term shifts in input mix brought about by the price changes.

Aspects of Costing Genebank Operations

We went to considerable lengths to ensure consistency in the scope and treatment of the cost data collected and assembled for this study. Doing so meant addressing several conceptual and practical issues.

Evolving protocols

During the period studied, most genebanks were restructuring and reorganizing their operations, with consequent changes in some of their conservation protocols. In many cases these changes were stimulated by the findings of the 1995 System-wide Genetic Resources Programme (SGRP) review of the centre genebanks (SGRP, 1996), and in some other cases they represented plans put into practice by individual centres. For example, one genebank was reconfiguring its storage space across crops to more efficiently manage the space; another was building new structures to accommodate expanded operations. Cost profiles during a transitional period can be quite different from the structure of costs when operations are being managed in a steady state. For this study we sought to compile and analyse the data for a 'representative' snapshot year, abstracting from abnormal aspects and assuming away technological changes when projecting these representative costs forward to simulate costs incurred in future years.

Jointness–divisibility

The genebank is but one of many programmes in a CG centre. Typically, some of the services required for operating a genebank are provided centrally and are shared with other programmes. For example, seed health testing units, field operation units or engineering units usually supply services to various programmes within a centre, thereby realizing scale economies and other efficiencies. A genebank operating as a stand-alone facility would have to secure each of these services independently, resulting in higher costs than those reported here. This study treats the costs of the shared operations as divisible among programmes; hence they are partially allocated to the genebank based on its share of service utilization. The costs of other centrally provided services (such as security, building maintenance and library) that cannot be allocated in this way are included as prorated parts of overhead costs.

The issue of jointness also arises within the genebank operation. When accessions are regenerated as a result of either low viability or low stock, the general practice is to regenerate enough seeds for both medium- and long-term storage, even though the main purpose of the regeneration is to replenish seed stocks in only one part of the storage facility. This study assumes that the regeneration is performed for both purposes and the total costs of regeneration are allocated equally between conservation and distribution functions. Similarly, when seeds are packed after cleaning and drying, all the packing for different purposes (for example, long- and medium-term storage, safety duplication, repatriation, distribution, and so on) is done at the same time. Again, this study assumes that the packing is divisible and allocates the packing costs to different operations according to the amount of material and labour required for each purpose.

Quality of operation

The genebank standards of FAO and the International Plant Genetic Resources Institute (IPGRI) lay out two sets of conservation standards. One is an 'acceptable standard' considered to be minimal but adequate, at least for the short term. The other is the 'preferred standard', which describes the basic conservation conditions (based on scientific criteria) that give a 'higher and thus safer standard' (FAO/IPGRI, 1994, p. 1). The funding realities are such that most CG genebanks have insufficient resources to satisfy the criteria required to meet the preferred standard. Thus genebank managers are forced to juggle priorities continually, meeting some aspects of the preferred standard for some parts of the collection, implementing acceptable standard for other aspects and, in some cases, making do with less than acceptable standards.

Meeting the preferred standard clearly costs more than maintaining the holding in an acceptable condition; hence taking cost data at face value is tricky. A comparatively high cost for a certain operation in one genebank does not necessarily imply that this operation is being achieved with less efficiency than the same operation at a lower-cost genebank. It might simply indicate a higher standard of operation. Because quality standards vary among centres and within centres over time, comparing costs on the premise that all else is equal can be quite misleading.

Capital costs

To estimate the annualized 'user cost' of capital, we compiled information on the purchase price of each capital item and combined the results with notions of the service profiles of each item and the real rate of interest. Past capital purchases were made on different dates, so they were inflated forward, using the most applicable price index series to express them in a set of base-year prices (assuming no technological change). We also assumed a 'one-hoss-shay' depreciation profile under which the capital good survives intact until the end of its life and then disappears all at once.[1] Annual depreciation costs are constant under this profile, so the annualized cost is easily calculated using the interest rate and service lives of each item (equation (3) in Appendix A).

Dynamic costs and life-cycle considerations

The costs of some operations such as storage are incurred annually, while the costs of other operations such as regeneration are incurred periodically – say, every 20–30 years – and the viability of a sample is tested every 5 years or so. Thus the conservation costs of a sample in any particular year depend on the time in storage and the status of the sample.

Figure 2.2 illustrates an example of the profile of conservation costs incurred during the life cycle of an accession from introduction, expressed in present-value terms with a positive discount rate. When an accession is newly introduced into a genebank at time zero, it is typically regenerated and tested for viability and health, and the costs of conservation in that year are especially high. During a normal year when an accession is simply held in storage (such as time t_A in Fig. 2.2), the conservation cost consists only of the long-term costs of storage. When an accession requires regeneration after failing a viability test, the costs in that year (time t_B in Fig. 2.2) are higher than the cost at time t_A. Year t_C represents a year in which a sample successfully passes a viability test and requires no regeneration. The present value of conserving an accession in perpetuity is obtained by summing all the areas (irrespective of their shading) of the bar graph in Fig. 2.2.

In the case studies, we calculated two types of costs: annual costs and in-perpetuity costs. The cost of conserving (and distributing) an accession for 1 year is the basis for estimating the overall annual expenditure of the whole operation. It depends on the crop in question and the state of the sample, including the time in storage and the time from last regeneration or viability test. The in-perpetuity cost is the basis for estimating the size of the conservation endowment fund (as discussed in Chapter 1). This cost also depends on a host of other factors made explicit in the chapters to follow, and is calculated by assuming: (i) no changes in the costs per accession over time in real (that is, inflation-adjusted) terms; and (ii) the maintenance of baseline conservation protocols throughout the entire period.

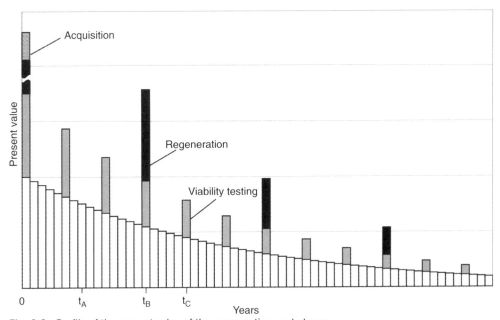

Fig. 2.2. Profile of the present value of the conservation cost stream.

Existing Studies on *Ex Situ* Conservation

Costs of conserving germplasm depend on many factors, such as crop type, conservation methods, natural environment and the institutional arrangements whereby germplasm are stored. Few studies, however, attempt to incorporate cost elements comprehensively in sufficient detail to address the cost implications of different conservation protocols for different crops in different environments. Some studies (for example, Burstin *et al.*, 1997; Virchow, 1999, 2003) adopt an indirect approach by collecting survey data. Subjective, inconsistent responses and excessive aggregation are identified as the main limitations of these survey-based approaches. To cope with these limitations, other studies collected data directly through on-site visits and interviews (for example, Epperson *et al.*, 1997; Pardey *et al.*, 1999, 2001), though at a significant cost in terms of time and effort.

Using an international survey conducted during 1995/96, Virchow (1999, 2003) collected national conservation expenditure data from 39 countries and estimated per-accession annual conservation expenditures for each country. The estimated expenditure ranges from $1 per accession in Austria to $1293 per accession in Egypt, and these estimates are used to compare the efficiencies of individual countries in conserving plant genetic resources. The limitations of that study include vaguely defined scope of the conservation activities, exceptionally aggregated data and country-dependent degrees of precision in the responses. The reported total expenditures cover both *in situ* and *ex situ* conservation efforts, while the per-accession cost estimates were developed using the number of accessions held only in *ex situ* collections.

A more focused survey-based study was conducted by Burstin *et al.* (1997), who examined the costs associated with conserving sexually and vegetatively propagated species in several French national genebank networks. Using survey data on the costs of storing, regenerating and characterizing each crop, together with the projected cycles of each operation, the authors calculated the annual and long-term costs of each operation. Different genebank managers, however, may have used different definitions or included different operational protocols; thus the corresponding cost data may not be comparable among crops. Another limitation of this study is that it did not discount the cost flows to net present value when calculating the costs of periodically performed operations.

Though not a costing study *per se*, SGRP (2000) reported the required expenditure to upgrade the CG genebank operations to international standards. In addition to their current annual operating budgets, the report estimated that an additional $20.8 million was required to upgrade all 11 CG genebanks.[2] Naturally, genebanks that fall far below the acceptable standard need more expenditure than those that are close to it. Thus, the size of the funding sought in this report reflected the status of genebank operations relative to the target level; it did not provide information on the current cost of conserving accessions along with the structural or managerial changes that would lead to improved cost efficiencies of the genebank system as a whole or its component parts.

The problems with survey-based approaches are evident. Even if the survey is carefully designed and explained, genebanks are so heterogeneous and complex that the data are often inconsistent, subject to the respondent's arbitrary interpretation. An alternative approach is to directly collect data on every cost element through on-site visits and interviews, and reconstruct the costs of genebank operation to obtain consistent and comparable estimates. Epperson *et al.* (1997) analysed the costs of conserving the cassava collection at CIAT in Colombia with this approach. They estimated the variable and fixed costs of conserving cassava germplasm in both the *in vitro* and field genebanks. While their study is a substantial improvement over previous survey-based studies, it considered only cost estimates for a single year without accounting for the different states of stored accessions and the dynamic aspects of conservation costs.

Notes

[1] A 'one-hoss-shay' profile is like a horse-drawn buggy from days of old that provides more or less the same services year in, year out until one year when it is no longer usable.

[2] In April 2003, at the time of completing this report, a total of $13.6 million had been earmarked for upgrading the 11 CG genebanks and the System-wide Information Network for Genetic Resources (SINGER) from World Bank funds to the CGIAR.

3 CIMMYT Genebank*

PHILIP G. PARDEY, BONWOO KOO, M. ERIC VAN DUSEN, BENT SKOVMAND, SUKETOSHI TABA AND BRIAN D. WRIGHT

History of the CIMMYT Genebank

Wheat collection

CIMMYT's present wheat collection was begun in about 1968. Throughout the 1970s CIMMYT's wheat holdings were essentially a working collection, preserving the parental material used in and the advanced lines coming from the breeding programme, including the material distributed through an international nursery system. No active acquisition programme existed; neither were there any systematic efforts to regenerate the holdings.

Beginning in the mid-1980s, the growth in the wheat collection accelerated as a consequence of increased political attention paid to (and hence resources made available for) the collection and conservation of plant genetic resources. From 1987 to 1997, the collection increased from 40,000 to 123,000 lines. The current wheat collection is a mixture of advanced breeding lines and parental germplasm from the CIMMYT breeding programmes, landrace collections from various regions of the world (principally Turkey, Pakistan, Iran and Mexico), and material provided from the collections or breeding programmes of other research agencies in other countries (especially North America, Japan, Denmark and the UK). The founding CIMMYT wheat collection contained mainly bread wheat, and was subsequently diversified through the addition of durum wheat, barley and triticale. The collection now consists of wheat at all stages of enhancement, from various wild and weedy species, through landraces and obsolete wheat cultivars to élite commercial cultivars.

*This chapter is a modified version of Pardey, P.G., Koo, B., Van Dusen, M.E., Skovmand, B., Taba, S. and Wright, B.D. (2001) Costing the conservation of genetic resources: CIMMYT's *ex situ* maize and wheat collection. *Crop Science* 41(4), 1286–1299.

The acquisition of varieties held in other *ex situ* collections is a significant means of growth in the collection. An example is a joint University of
California, Davis–CIMMYT project conducted during 1988/89 that rescued
more than 3000 triticale lines (wheat–rye crosses) from the collections of
three prominent North American triticale breeders (Furman *et al.*, 1997).
Every year the collection grows further with the addition of advanced
breeding lines from CIMMYT's crop-improvement programme. Prominent
among these are the sets of advanced wheat breeding lines (and improved
barley varieties) released for trials and evaluation around the world in
CIMMYT's International Nurseries programme.

In addition, field collection expeditions are undertaken by the CIMMYT
genebank staff to acquire germplasm that may be endangered or be deemed
under-represented in the existing collection or of special interest for its
breeding potential. During the period 1992–1997, CIMMYT added about
10,000 rye and barley accessions (from about 300 locations throughout
Mexico) to its holdings. This material is now being characterized by
CIMMYT, and a complete set of the collection has been repatriated to the
national programme. Table 3.1 summarizes the past changes and current
status of the number and type of wheat accessions held at CIMMYT.

Maize collection

The CIMMYT maize holdings are based on a collection first assembled as
part of the joint Rockefeller Foundation–Government of Mexico programme initiated in 1943 to improve the productivity of basic food crops in
Mexico. An Office of Special Studies (OSS) was formed within the Mexican
Ministry of Agriculture to carry out this programme of research, and a
Mexican seed bank was established in 1944. By 1947 its maize collection
had grown to more than 2000 samples (mainly landraces). With the forma-
tion of a US National Academy of Sciences (NAS)–National Research
Council (NRC) initiative in the early 1950s, additional samples of maize
were assembled from expeditions throughout Latin America, the USA and
Canada. The NAS–NRC effort collected nearly 11,000 samples. Seeds from
Mexico, Central America and the Caribbean region were stored in the
Mexican genebank in Chapingo, maintained by OSS and operated by them
until 1959.

The closure of the OSS and the subsequent transfer of its maize holdings to the newly formed national research agency coincided with the
launch of the Inter-American Maize Program. This programme, a joint ven-
ture between the Government of Mexico and the Rockefeller Foundation,
regenerated and duplicated the entire collection in the Instituto Nacional de
Investigaciones Agrícolas (INIA). The programme also regenerated part of
the Latin American NAS–NRC collection, which was shipped from the
NSSL facility in Fort Collins to Mexico and formed the basis of the
CIMMYT maize collection. CIMMYT participated in various maize collection expeditions in Mexico and the Andean region in the late 1960s. Maize

Table 3.1. CIMMYT genebank holdings, 1970–1997. (From FAO, 1996, and CIMMYT genebank data files.)

Crop	Number of accessions				Origin of 1997 holdings[a]	
	1970	1980[b]	1990	1997	From CIMMYT breeding programme (%)	Others (%)
Wheat collection						
Bread wheat	n/a	4,505	42,881	71,171	60	40
Durum wheat	n/a	2,140	11,689	15,490	60	40
Triticale	n/a	2,240	8,576	15,200	85	15
Barley	n/a	2,096	7,918	9,084	75	25
Rye	n/a	–	33	202	25	75
Primitive and wild	n/a	–	3,934	11,794	0	100
Total	**n/a**	**10,981**	**75,031**	**122,941**		
Maize collection						
Zea mays	4,612	9,869	10,364	17,000	4	96
Tripsacum	7	39[c]	39[c]	181[c]	100	0
Teosinte	36	124	130	162	100	0
Total	**4,655**	**10,032**	**10,533**	**17,343**		

[a]Wheat data are approximate shares.
[b]Wheat data for 1980 are estimates.
[c]Additional collections are held by CIMMYT, but not formally as part of the bank inventory.
n/a, not available.

collection expeditions sponsored by IPGRI's precursor, the International Board of Plant Genetic Resources (IBPGR), throughout Latin America, southern Europe and Asia got under way in 1975 (Reid and Konopka, 1988). This IPGRI-related work continued through to 1985, by which time a further 1500 samples had been added to the CIMMYT collection.

The CIMMYT holdings grew at a more rapid rate thereafter, to its present size of more than 17,000 accessions. This accelerated growth was largely a consequence of the Special Cooperative Agreement to regenerate Latin American maize germplasm, which involved 13 countries during 1992–1996. By 1996 a total of 6736 accessions had been regenerated, and backup samples were shipped to NSSL and CIMMYT for long-term storage. CIMMYT also recorded characterization data for the regenerated samples (Taba and Eberhart, 1997). In addition, the Latin American Maize Project (LAMP), funded by Pioneer Hi-Bred International and coordinated by the US Department of Agriculture (USDA) during 1987–1996, evaluated about 12,000 Latin American accessions (Salhuana *et al.*, 1998). Another major source of new genebank accessions is the CIMMYT maize-breeding programme, from which samples of élite experimental varieties, source populations and inbred lines are obtained.

Aside from *Zea mays* (cultivated maize), the CIMMYT collection includes two other species important to maize breeders. During 1989–1992, CIMMYT collected 2500 samples as cuttings from 158 populations of the perennial genus *Tripsacum* located throughout Mexico. About 150 of these samples have been established in a living base collection at the CIMMYT field station in Tlaltizapan, Morelos. This material is being used in a joint Institut de Recherche pour le Développement (IRD) (France)–CIMMYT undertaking that applies new molecular tools to study the transfer of apomixis from *Tripsacum* to *Z. mays* (Berthaud *et al.*, 1997). CIMMYT also maintains a collection of teosinte, the closest wild relative of maize. Because teosinte outcrosses with maize or other teosinte accessions, multiplication and conservation of these plants must occur in isolation on experimental plots, using open pollination among more than 100 plants if possible (Taba, 1997).

Genebank facility

CIMMYT's early operations were geared almost exclusively to improving wheat and maize yields, based largely on the development of improved varieties. The institute's germplasm holdings reflected that crop-breeding focus. From 1966 to 1971, the CIMMYT maize collection (developed by regenerating material as part of the Inter-American Maize Program) was housed in refrigerated storage facilities in the basement of the soil science building at the National School of Agriculture, Chapingo. In 1971 a new seed-storage facility was completed at CIMMYT in El Batan, Texcoco, and the collection was subsequently transferred there. The facility consisted of two 145 m^3, refrigerated chambers held at 0°C, but in 1984 was refitted to provide one chamber for the long-term storage of a base maize collection held at $-18°C$. The other chamber was retained for an active maize collection. By the late 1980s, the storage space was almost filled to capacity.

CIMMYT's wheat holdings were initially stored in small, paper packets held in freezer chests. In 1981 the wheat collection was moved to a newly constructed 1500 m^2 facility with four refrigerated chambers. Two chambers, with a combined capacity of 90,000 accessions, were maintained at 4–5°C for an active collection of germplasm, and two larger chambers, with a combined capacity of 180,000 accessions, were kept at about $-2°C$ for medium-term storage of a base collection of CIMMYT's research products. However, during the 1980s, CIMMYT's objectives gradually broadened to include germplasm conservation (specifically the development and maintenance of a comprehensive bread-wheat and triticale collection), and it became necessary to develop a suitable low-temperature, low-moisture facility to house this new base collection over the longer term.

In October 1995, construction of a new genebank facility, financed by the Japanese government, was commenced, and the main construction phase was completed by May 1996.[1] The maize and wheat collections were consolidated into a single facility, with advanced technology for medium- and long-term storage. The main structure of the new genebank facility consists of a two-storey, fortified-concrete bunker, built to withstand most conceivable natural or other disasters. The climate is controlled to precise temperature and humidity specifications, and the facility is equipped with alarms, security measures and a backup power supply. The upper (ground) level of the storage rooms houses the active collection, held at just below freezing-point $(-3°C)$ and 25–30% relative humidity. This constitutes the 'working' part of the bank, from which seed requests by CIMMYT and other scientists are filled. The lower (below-ground) level consists of the base collection stored at $-18°C$, primarily for long-term storage. The seeds are stored on movable shelves to optimize use of the available space.

Costing the CIMMYT Genebank

Capital input costs

A breakdown of the capital costs related to the genebank facility and the costs of the equipment used in CIMMYT's genebank operation is provided in Table 3.2. Complementing the storage facility are rooms for cleaning, sorting and packing seeds destined for storage at CIMMYT or shipment elsewhere, drying rooms, various work rooms, offices and a seed laboratory used for germination testing that is shared between the maize and wheat programmes. The genebank is also serviced by a backup power-generation unit. Much, but not all, of the backup power unit is dedicated to the genebank; about 80% of this cost was allocated to the genebank, based on consultation with CIMMYT's plant managers. Costs that were common to storing the maize and wheat collections were allocated equally to each crop.

Table 3.2. Capital input costs (US$, 1996 prices) at the CIMMYT genebank.

Cost category	Service life (years)	Replacement cost		Annualized cost[a]	
		Wheat	Maize	Wheat	Maize
Medium-term storage		**277,765**	**311,835**	**19,979**	**22,076**
Storage facility	40	174,051	174,051	8,455	8,455
Storage equipment	10	81,979	81,979	9,718	9,718
Backup power system	10	8,205	8,205	973	973
Seed container	25	13,530	47,600	833	2,930
Long-term storage		**274,485**	**271,335**	**19,752**	**19,611**
Storage facility	40	174,051	174,051	8,455	8,455
Storage equipment	10	81,979	81,979	9,718	9,718
Backup power system	10	8,205	8,205	973	973
Vacuum sealer	10	2,000	2,000	237	237
Seed container	50	8,250	5,100	369	228
Germination testing		**12,650**	**12,650**	**1,052**	**1,052**
Germination testing facility	40	6,400	6,400	311	311
Germination chamber	10	6,000	6,000	711	711
Other equipment	10	250	250	30	30
Regeneration		**206,500**	**109,965**	**23,464**	**12,561**
Screenhouse	10	112,000	–	13,277	–
Vernalizer	10	12,000	–	1,423	–
Seed-cleaning equipment	10	–	4,465	–	529
Seed-drying equipment	10	25,000	35,000	2,964	4,149
Seed-processing facility	40	30,000	30,000	1,457	1,457
Seed-processing equipment	10	1,500	1,500	178	178
Vehicle	7	26,000	39,000	4,165	6,248
Seed-health testing		**14,429**	**9,171**	**1,659**	**1,052**
Seed-health testing facility	40	1,296	864	63	42
Greenhouse	10	1,080	720	128	85
Lab/office equipment	10	10,445	6,963	1,238	825
Jacuzzi equipment	10	672	–	80	–
Vehicle	7	936	624	150	100
General capital		**34,800**	**34,800**	**2,390**	**2,390**
General facility	40	24,800	24,800	1,205	1,205
Office equipment	10	10,000	10,000	1,185	1,185
Total capital costs		**820,629**	**749,756**	**68,296**	**58,742**

Note: See Appendix B for further details.
[a]Calculated at a 4% baseline interest rate using equation (3) in Appendix A.

Table 3.2 also lists the capital costs associated with seed-health operations, part of which are prorated to the genebank. The balance is charged to
CIMMYT's breeding and international nursery-trials operations, which are
also serviced by the seed-health unit. Some custom-built Jacuzzi equipment
is used to clean seed for Karnal bunt (*Tilletia indica*) disease, and such seed
is shipped overseas as part of CIMMYT's international wheat nursery programme or in response to requests for seed from the genebank. Thus, its use
was prorated between two programmes accordingly.

Annual operating costs

Storing seeds

Maintaining the storage areas in the genebank at precise and stable low temperatures and low relative humidity is a costly exercise. The variable costs of controlling the climate in the CIMMYT facility include the cost of electricity to run the compressors, dehumidifiers and fans; the costs of maintaining this equipment; and the related costs of operating an emergency backup power plant. Allocating these types of costs to the germplasm facility is difficult as they represent only part of the overall costs involved in operating the institute's physical plant. To arrive at the estimates in Table 3.3, we directly costed the energy required to maintain the genebank at its specified climate, and also estimated the costs of a routine schedule of maintenance on the climate-control equipment and backup power-generation unit.

The information management costs in Table 3.3 represent the costs of creating, updating and managing the various databases used in the genebank operation. This includes the cost of developing software by CIMMYT's computer-support staff, so we removed this expense from the general CIMMYT overhead rate to avoid double counting. The 'general management' category includes the costs of the genebank managers and technical staff and other general genebank costs. The conservation of genetic resources is a primary rationale for maintaining a genebank separate from the working collections maintained by breeders. From this perspective these expenses represent a rather lumpy set of costs that were prorated among the various conservation and dissemination functions identified in the tables to follow.

Germination testing

Stored seeds gradually lose their viability through ageing and so their germination rates must be checked periodically. For wheat, the monitoring and regeneration procedures followed by CIMMYT begin with a germination test when processing seed on its first entry to the genebank or after its last regeneration. A sample of the seed from each accession is placed in a germination chamber for 5 days and checked to determine its viability: the accession undergoes a cycle of regeneration if its germination rate falls below 85%. If the sample satisfies the viability criterion, it is retested at a later time. A computer program is used to sample from the active collection for germination testing, selecting a number of 5-year-old accessions, more 10-year-old seeds, even more 20-year-old seeds, and so on. For now, the maize genebank also samples from the active collection for germination testing (beginning with the oldest seed first and working forward), restoring both the active and base collection if the sample fails to germinate satisfactorily. Eventually, this procedure (rotating through the collection from the oldest to the youngest samples) settles at a 5-year cycle.

A large share of the costs in assessing viability consists of the costs of the labour used to carry out the tests, but additional costs (including the costs of establishing and running a suitable laboratory with germination

Table 3.3. Annual operating costs (US$, 1996 prices) of conservation and distribution at the CIMMYT genebank.

Cost category	Wheat				Maize			
	Labour	Non-labour	Subtotal	Capital	Labour	Non-labour	Subtotal	Capital
Acquisition	**10,582**	**2,907**	**13,489**	**995**	**5,993**	**1,621**	**7,614**	**632**
Seed-health testing	6,148	1,949	8,097	–	4,187	1,327	5,514	–
Introductory planting	1,996	431	2,427	–	–	–	–	–
Seed handling	520	–	520	–	720	–	720	–
Overheads	1,918	527	2,445	–	1,086	294	1,380	–
(Number of accessions)			(5,800)				(1,580)	
Medium-term storage	**10,467**	**1,962**	**12,429**	**19,979**	**10,467**	**1,962**	**12,429**	**22,076**
Storage management	6,230	–	6,230	–	6,230	–	6,230	–
Climate control	2,340	1,606	3,946	–	2,340	1,606	3,946	–
Overheads	1,897	356	2,253	–	1,897	356	2,253	–
(Number of accessions)			(123,000)				(17,000)	
Long-term storage	**7,424**	**3,312**	**10,736**	**19,753**	**7,424**	**3,312**	**10,736**	**19,612**
Storage management	3,738	–	3,738	–	3,738	–	3,738	–
Climate control	2,340	2,712	5,052	–	2,340	2,712	5,052	–
Overheads	1,346	600	1,946	–	1,346	600	1,946	–
(Number of accessions)			(75,000)				(17,000)	
Germination testing	**6,537**	**244**	**6,781**	**1,052**	**4,314**	**122**	**4,436**	**1,052**
Germination testing	5,352	200	5,552	–	3,532	100	3,632	–
Overheads	1,185	44	1,229	–	782	22	804	–
(Number of accessions)			(12,000)				(3,400)	
Dissemination	**31,768**	**5,171**	**36,939**	**664**	**16,776**	**8,402**	**25,178**	**421**
Dissemination management	22,428	–	22,428	–	11,214	–	11,214	–
Seed-health testing	2,753	984	3,737	–	1,884	673	2,557	–
Seed treatment	90	490	580	–	–	–	–	–
Packing and shipping	738	2,760	3,498	–	637	6,206	6,843	–
Overheads	5,759	937	6,696	–	3,041	1,523	4,564	–
(Number of accessions)			(14,220)				(3,680)	

Duplication	5,822	5,186	11,008	—	1,887	4,169	6,056	—
Packing and shipping	4,767	4,246	9,013	—	1,545	3,413	4,958	—
Overheads	1,055	940	1,995	—	342	756	1,098	—
(Number of accessions)			*(35,000)*				*(2,230)*	
Information management	22,900	611	23,511	—	28,725	611	29,336	—
Database management	14,280	—	14,280	—	17,400	—	17,400	—
Catalogue management	4,469	—	4,469	—	6,118	—	6,118	—
Other expenses	—	500	500	—	—	500	500	—
Overheads	4,151	111	4,262	—	5,207	111	5,318	—
General management	40,208	13,832	54,041	2,390	55,427	15,542	70,969	2,390
Managerial staff	32,920	—	32,920	—	45,380	—	45,380	—
Computers	—	4,900	4,900	—	—	6,300	6,300	—
Electricity	—	1,425	1,425	—	—	1,425	1,425	—
Other expenses	—	5,000	5,000	—	—	5,000	5,000	—
Overheads	7,288	2,507	9,796	—	10,047	2,817	12,864	—
Total operating costs	135,708	33,225	168,934	44,833	131,013	35,741	166,754	46,183

Note: See Appendix B for further details.

chambers) must be factored in as well. The operational costs associated with germination testing are reported in Table 3.3, along with the respective annualized capital cost from Table 3.2.

Regeneration

A principal challenge in managing the regeneration of an *ex situ* collection is to minimize the prospects of genetic drift, thereby maintaining a collection whose genetic make-up matches that of the original holdings as closely as possible. Genetic drift involves the loss of alleles (that is, genetic content) from one regeneration cycle to another. The frequency of regeneration cycles can be increased to maintain sample size, but the regenerative process itself must be carefully managed to minimize genetic drift. For example, in an open-pollinating crop like maize, if some seeds in an accession have the propensity for higher pollen production than others, hand-pollination may be necessary to prevent drift towards the higher-pollen characteristic. Moreover, genetic drift may be exacerbated if samples are regenerated under conditions of soil, chemical inputs or daylight that differ markedly from the native ecology. This is generally more of a concern when regenerating wild relatives and some landraces specifically adapted to their growing environments than when regenerating more advanced breeding lines and improved cultivars.

Wheat accessions are now normally regenerated in a screenhouse at El Batan or in Mexicali. The screenhouse facility enables regeneration to proceed year-round under controlled and protected conditions, with up to three cycles per year at staggered times to spread the use of labour. In 1996, the sample year for this study, an exceptionally large number of accessions were regenerated to deal with potential Karnal bunt problems when transferring materials from the old to the new storage facility opened that year. Seed samples were first prepared in special plots at the El Batan field stations (and sprayed with fungicides every 10 days to prevent Karnal bunt infestation), and then flown to Mexicali, in the state of Baja California Norte, where they were sown out in 1 m rows to scale up the size of the sample to 500 g. The peak labour requirements in the regeneration process occur at the time of harvest and during the completion of field books, wherein various morphological and physiological traits for each accession are recorded.

Most of CIMMYT's maize accessions obtained from tropical maize-growing areas of low and intermediate elevations are regenerated at Tlaltizapan, Morelos, while El Batan is used for germplasm obtained from the tropical highlands. Maize uses 2.5 ha in Tlaltizapan for two cycles per year and 1.5 ha in El Batan. A minimum of 16 5-m rows are required to regenerate a maize accession, but we based our calculations on a 20-row standard to account for failed regeneration. Since there are approximately 2000 rows/ha (100 accessions), it requires a total of 6.5 ha to regenerate 650 maize accessions.

Table 3.4 reports the varying amount of inputs used, such as irrigation, agrochemicals (including fertilizers) and management time, according to seasonal and other factors. Given that many of these costs are not explicitly

Table 3.4. Annual regeneration costs (US$, 1996 prices) at the CIMMYT genebank.

Cost category	Wheat				Maize			
	Labour	Non-labour	Subtotal	Capital	Labour	Non-labour	Subtotal	Capital
Regeneration	**74,498**	**15,055**	**89,553**	**23,464**	**81,797**	**21,646**	**103,443**	**12,562**
Field operation	**56,250**	**11,833**	**68,083**	**18,865**	**62,548**	**13,356**	**75,904**	**6,248**
Field management	24,920	–	24,920	4,165	24,920	–	24,920	6,248
Seed preparation	520	220	740	–	520	7	527	–
At El Batan								
Planting	2,340	242	2,582	14,700	1,040	–	1,040	–
Plant maintenance	3,006	1,480	4,486	–	500	1,110	1,610	–
Pollination	–	–	–	–	2,080	1,459	3,539	–
Harvesting	4,160	330	4,490	–	3,120	15	3,135	–
At Mexicali/Tlaltizapan								
Planting	1,404	–	1,404	–	1,560	–	1,560	–
Plant maintenance	1,800	3,068	4,868	–	2,130	2,915	5,045	–
Pollination	–	–	–	–	7,540	4,864	12,404	–
Harvesting	7,904	330	8,234	–	7,800	15	7,815	–
Transportation	–	4,018	4,018	–	–	550	550	–
Overheads	10,196	2,145	12,341	–	11,338	2,421	13,759	–
Seed processing	**18,248**	**3,222**	**21,470**	**4,599**	**19,249**	**8,290**	**27,539**	**6,314**
Processing management	6,230	–	6,230	1,635	6,230	–	6,230	1,635
Seed cleaning/drying	7,280	2,638	9,918	2,964	9,360	6,787	16,147	4,679
Medium-term packing	715	–	715	–	85	–	85	–
Long-term packing	715	–	715	–	85	–	85	–
Overheads	3,308	584	3,892	–	3,489	1,503	4,992	–
(Number of accessions)			*(22,000)*				*(650)*	

Note: See Appendix B for further details.

itemized in CIMMYT's accounting system, we first estimated the typical quantity of each input for regeneration and then priced each item accordingly to derive corresponding input costs. We used these benchmark field costs to estimate the overall costs involved in regenerating an accession of wheat and one of maize, taking care to adjust the figures to reflect cost differentials arising from differences in the seed density, seed volume and reproductive aspects of each crop. The labour costs for maize are much higher than for wheat because of the hand-pollination required for each maize plant. Regenerating maize also involves additional costs associated with the glassine and pollination bags used to control pollination. Wheat, on the other hand, uses a screenhouse, thereby pushing up its associated capital costs.

Processing seed accessions for storage

Prior to this study, the wheat programme routinely regenerated incoming accessions before introducing them to the genebank, whereas the maize programme generally did not. If regeneration is performed, processing a new introduction to the genebank is much like regenerating an existing accession but involves certain additional treatments. The introduced seed is inspected thoroughly upon arrival to screen for any known or suspected seed-health problems, which if found result in the seed being burned. Wheat and maize seeds are then deep-frozen until planting out to kill any insects. The first regeneration is performed on specially quarantined introduction plots under stringent pest-control procedures. The seed-health unit inspects the plants during this process as well as the resulting seed. After harvesting from the introduction plots, maize seeds are formed into bulk samples and added directly to the genebank. Wheat seeds planted out at El Batan undergo a further round of regeneration in the screenhouse to improve the quality of the seed in readiness for storage. In addition to the seed-health aspects, various characterization and data-entry activities are performed before an accession is finally added to the collection.

It typically takes much more time to manually clean, sort and inspect maize seeds than it does wheat seeds: each ear of maize must be sorted by hand to remove broken or diseased seeds. Although wheat seeds are intrinsically easier to handle, they require comparatively more attention to aspects of seed health, as discussed above and in more detail below. Both maize and wheat accessions require a similar amount of labour to record relevant data in field books, but the higher planting density for (and smaller growth habit of) wheat offers some time efficiencies compared with maize.

Each wheat and maize accession is stored at CIMMYT headquarters in two sets of containers – one going to the active collection and the other to long-term storage in the base collection. Each wheat accession, whether for active or long-term storage, is stored in an aluminium bag that costs 11 cents; each maize accession held in the active collection is sealed in a plastic bucket that costs $2.80, and each accession stored in the base collection is placed in two aluminium bags that cost 15 cents each (the bags used for maize are larger versions of the bags used for wheat). In addition, a sample of each accession (10 g of wheat seed and 1.5 kg of maize seed) is prepared for backup storage in the NSSL in Fort Collins, Colorado (discussed further under 'Seed dissemination and duplication', below).

Seed health
All newly introduced material is subject to seed-health checks before being included in the genebank. The health of all outgoing seed must also be certified and our cost schedules reflect this requirement. We took care, however, not to double-count health costs; in all but exceptional cases the checks done at the time of introducing or regenerating maize seed suffice for subsequent shipments made from the collection. Wheat seeds are checked when first introduced and again at the time samples are packaged for shipment. At CIMMYT, most of the relevant seed-health activity and the associated costs are the responsibility of CIMMYT's seed-health unit. The labour and other operational costs for the genebank operation are included in Table 3.3 within the acquisition and dissemination costs – though only part of the seed-health operation relates to accessions coming into and being shipped from the genebank, so only part of the overall seed-health costs are included.

Some seed-health costs are incurred directly by the genebank. The general operation of a well-managed genebank involves periodic checking for ambient (airborne) spores, monitoring the cleanliness of the machinery used in processing the seed and precautionary measures to eliminate possible contamination, which at CIMMYT involves the daily washing of all walls and floors with bleach in areas where seeds are processed. The efforts to deal with Karnal bunt also had cost consequences for the genebank. Karnal bunt is not a particularly virulent or economically important disease for wheat, but its presence does limit the acceptability of seed that is infected or contaminated by the fungus by numerous national quarantine agencies (Fuentes-Davila, 1996; Beattie and Biggerstaff, 1999).

CIMMYT's troubles with this disease stem from an infestation of Karnal bunt in the CIMMYT fields at Ciudad Obregon, Sonora, that were routinely used by the genebank prior to 1987. Although the Sonora fields are ideal in many respects for regenerating seeds, they are no longer used because of the Karnal bunt problem. Instead, wheat seeds are now multiplied in clean plots at El Batan, checked for spores in bulked samples after being passed through chlorine disinfection, regenerated at Mexicali and shipped back to El Batan in sealed containers. To facilitate large-scale disinfestation for Karnal bunt a Jacuzzi-like system for cleaning wheat seeds was developed, as mentioned above. The costs of dealing with the contamination involve additional regeneration costs, specialized shipping procedures and related phytosanitary certification costs, increased chemical applications and increased seed-health monitoring costs. These costs are incorporated into the estimates provided in Table 3.3.

Seed dissemination and duplication
Distribution from the genebank takes various forms. Some material is used by genebank personnel for characterization or evaluation purposes. Other

material is distributed in response to individual requests from breeders, plant pathologists and others at CIMMYT or elsewhere. Seed is also sent to other genebank facilities, often in the context of CIMMYT's joint collection and conservation work with developing-country national agricultural research systems (NARS). Responding to such a diverse set of seed requests entails the costs of determining which seeds are most suitable to fill the request; assembling, treating and packaging the samples to be sent; and then shipping them.

Another set of costs is sensitive to the number of shipments made (as distinct from the number of accessions shipped), as well as the size and destination of each shipment. Relatedly, each shipment outside Mexico is subject to phytosanitary controls and this certification process is a reasonably time-intensive and costly undertaking. Aside from the cost of the certificates themselves (payable to the Mexican government), it draws on the time of staff in CIMMYT's seed-health unit and the genebank to prepare the necessary documentation and arrange for the shipment itself. In addition, shipments of seed from CIMMYT must be accompanied by a Material Transfer Agreement (MTA) that assigns use rights to the seed, and this documentation must be developed, tracked and logged.

Table 3.5 summarizes the shipments of germplasm made from the genebank since 1987. It also provides information on the number of shipments, as distinct from the number of accessions shipped. About 75% of the CIMMYT wheat shipments go to the crop-improvement programme and about 25% for evaluation activities managed by the genebank programme. Shipments from the genebank to the rest of the world are made on request, and so can vary substantially from year to year. For instance, the exceptionally large wheat shipments in 1987 reflected significant requests from India to aid their efforts to find resistance to Karnal bunt, and the large developing-country shipments in 1998 were due to the repatriation of material (9811 accessions in total) collected throughout Mexico to the national programme. More wheat than maize accessions are shipped abroad, and there are fewer wheat shipments per year. Thus the average number of accessions per shipment is 220 in the case of wheat compared with just 34 for maize.

The new storage facilities at CIMMYT are designed to withstand major natural catastrophes, and backup power generation, climate control and general operating procedures are also in place to minimize the chance of damage to or loss of the collection. As an additional safety precaution, much of the CIMMYT wheat and maize collections is duplicated at other locations. By 1997, about four-fifths of the base maize collection and approximately half of the base wheat collection were held at the NSSL mentioned earlier. CIMMYT has formal agreements for the storage of a duplicate collection with the NSSL for both maize and wheat and with ICARDA for wheat. Parts of the collection are also backed up less formally at other sites. The National Institute of Agrobiological Resources (NIAR) in Japan and the Australian Winter Cereals Collection (AWCC) hold significant parts of the CIMMYT wheat collection.

Table 3.5. Dissemination of germplasm from the CIMMYT genebank, 1987–1998.

Activity/crop	1987	1988	1989	1990	1991	1992	1993	1994	1995	1996	1997	1998
Number of samples[a]												
Wheat												
CIMMYT	2,764	1,690	4,928	940	4,042	2,278	6,333	1,026	2,944	12,890	8,624	2,652
Developing countries	9,287	288	2,547	680	324	561	584	3,793	229	133	542	11,601
Developed countries	195	92	2,269	490	21	115	1,160	703	101	1,200	1,822	1,003
Total	12,246	2,070	9,744	2,110	4,387	2,954	8,077	5,522	3,274	14,223	10,988	15,256
Maize												
CIMMYT	2,400	4,341	5,093	3,450	2,231	1,970	3,740	3,039	2,542	2,776 (2,607)	1,678 (1,574)	1,599 (883)
Developing countries	1,667	1,489	1,238	1,103	508	536	818	717	264	803	686	3,109
Developed countries	447	587	1,378	687	117	710	1,813	637	532	106	234	354
Total	4,514	6,417	7,709	5,240	2,856	3,216	6,371	4,393	3,338	3,685	2,598	5,062
Number of shipments												
Wheat												
CIMMYT	21	23	41	38	19	18	14	8	7	9	11	24
Developing countries	25	13	28	12	5	12	2	10	2	14	12	16
Developed countries	12	11	10	6	5	6	3	14	4	8	11	13
Total	58	47	79	56	29	36	19	32	13	31	34	53
Maize												
CIMMYT	27	48	47	46	27	25	37	39	34	28	26	28
Developing countries	32	20	48	21	22	25	22	23	17	28	30	50
Developed countries	19	29	17	20	16	15	21	18	13	13	19	16
Total	78	97	112	87	65	65	80	80	64	69	75	94

[a]The number of samples shipped from the CIMMYT genebank to CIMMYT include material sent to breeders, plant pathologists and others involved in the centre's crop-improvement programme as well as material destined for evaluation trials run by genebank personnel. The figures in parentheses indicate the number of maize accessions shipped to genebank personnel for evaluation purposes.

The backup collections for wheat are shipped and stored as 'black boxes': a 10 g wheat accession (around 350 seeds) is prepared, labelled and packed in aluminium-foil bags and then consolidated into cardboard boxes, containing up to 400 accessions each. The boxes are air-freighted to the backup facility where they are stored. The expenses of preparing the samples and packing each box are included in our costing calculations. Freight costs from El Batan to Fort Collins for the last shipment of black boxes in 1996 totalled $342 for 35,000 duplicates. Wheat duplicates are accumulated and shipped periodically to save shipping costs.

CIMMYT's maize holdings are duplicated and stored as an integral part of the NSSL collection rather than via the black-box method. New introductions and regenerated accessions are shipped to NSSL annually, and about 80% of the CIMMYT maize collection was backed up at NSSL by 1996. Between 1500 and 2000 accessions are shipped each year in cheesecloth bags after the regenerated seed is dried. The NSSL repack and store the accessions in aluminium bags. CIMMYT identity numbers are entered into NSSL's data-management system, along with information on the amount of seed in storage and its germination status.

Data and information management

Fundamental to the genebank is the management of the information that describes each accession. However, operationally (and for costing purposes) it is difficult to separate data and information used in the effective management of genebanks from the data that are generated by and facilitate the breeding programme at CIMMYT and elsewhere in the world. Some of the data serve multiple purposes. Significant benefits accrue for both seed conservation and breeding through the use of standardized accession ID numbers, uniform protocols for recording and reporting performance evaluation data (whether collected for genebank regeneration or under international evaluation trials), and compatible software for recording, storing, retrieving and analysing such data.

Routine operations of the genebank include entry of passport data, detailing the source and origin of the seed, at the time the accession first enters the collection; processing field observations collected under trials on new accessions and subsequent regenerations; and maintaining the database that tracks the storage location, time in storage, seed viability history and stock levels of each accession. Barcode labelling of each maize accession in the genebank is being introduced to streamline this process.

Genebank management systems are part of a broader effort at CIMMYT to improve the information base concerning the centre's extensive maize and wheat holdings. Several developments have occurred in recent years:

1. A new system called the Genetic Resources Information Package (GRIP) has been developed with combined project funding from Australia and core funding from CIMMYT. The information entered into this system has also been incorporated into the System-wide Information Network for Genetic Resources (SINGER).

2. The Wheat Genebank Management System has been incorporated into the International Wheat Information System (IWIS), a computer database system that integrates information from nursery trials through to pedigree information and is able to trace lineages of advanced breeding lines.

3. The Maize Germplasm Bank Management System (MZBANK) was also updated through participation in the SINGER project.

4. LAMP has produced a CD-ROM containing passport information for all of CIMMYT's maize-bank accessions, including the original collection and regenerations from ongoing LAMP collaboration.

Economic Analysis

Representative annual costs of genebank operation

Table 3.6 consolidates data presented in previous tables to provide an overview of the total variable and capital costs under various activities and cost components, along with the corresponding average cost per accession. The first column reports the number of treated or stored accessions implicit in each of the capital, quasi-fixed capital and variable cost totals listed in columns 3, 4 and 5, respectively. Taking these figures at face value, the total cost of conserving and distributing CIMMYT's wheat and maize collections in 1996 is $655,725 – $326,786 for wheat and $328,939 for maize.

However, some of the genebank operations represent atypical activity levels during the survey year, and we reverted to a more typical set of accession numbers in each operation to derive a more representative estimate of the annual cost totals.[2] Figure 3.1 breaks the adjusted annual operating cost of $543,089 into the various cost classes. About 65% of the annual cost of the genebank operation involves labour (41% for scientific quasi-fixed labour and 24% for non-technical labour) inputs, with the remaining costs divided between operational costs (13%) and the annualized cost of capital (22%). When expressed in annualized terms, the capital costs of running the CIMMYT genebank are not especially suggestive of a capital-intensive operation. Indeed, like the crop-improvement research it supports, maintaining a genebank is a labour-intensive undertaking and carries a significant, recurrent, 'overheads' cost through the labour required to manage the bank and regenerate and otherwise maintain the viability of the collection. About two-thirds of these labour costs (representing the cost of the senior scientific and technical staff) are lumpy and best treated as a quasi-fixed input.

The structure of the costs differs significantly between wheat and maize. While the annual cost of physical capital inputs differs little between CIMMYT's wheat and maize operations (and to a lesser extent the cost of lumpy labour inputs), there exist substantial differences in the structure of their variable costs. The maize programme spends considerably more than the wheat programme each year on regenerating its holdings. Indeed, regenerating seed accounts for 58% of the variable costs for maize and only

Table 3.6. Annual total (US$, 1996 prices) and average (US$ per accession, 1996 prices) costs of each operation at the CIMMYT genebank.

Cost category	Number of accessions	Total capital costs	Total quasi-fixed costs[a]	Total variable costs	Average capital costs	Average quasi-fixed costs	Average variable costs
Wheat		**68,299**	**121,648**	**136,839**	**2.02**	**6.87**	**7.68**
Medium-term storage (under full capacity)[b]	123,000	20,338	12,175	11,887	0.17	0.10	0.10
					0.05	0.03	0.10
Long-term storage (under full capacity)	75,000	19,873	6,087	8,526	0.26	0.08	0.11
					0.05	0.02	0.11
Acquisition	5,800	1,115	6,707	10,659	0.19	1.16	1.84
Germination testing	12,000	1,291	6,087	8,449	0.11	0.51	0.70
Regeneration	22,000	24,301	48,700	67,996	1.10	2.21	3.09
Safety duplication	35,000	120	4,566	10,320	0.00	0.13	0.29
Dissemination	14,200	1,261	37,326	19,002	0.09	2.63	1.34
Maize		**58,746**	**119,260**	**150,933**	**24.90**	**99.58**	**148.60**
Medium-term storage (under full capacity)	17,000	22,435	14,458	13,017	1.32	0.85	0.77
					0.33	0.22	0.77
Long-term storage (under full capacity)	17,000	19,732	6,848	8,902	1.16	0.40	0.52
					0.29	0.10	0.52
Acquisition	1,580	751	5,814	6,815	0.48	3.68	4.31
Germination testing	3,400	1,291	7,609	6,857	0.38	2.24	2.02
Regeneration	650	13,398	54,026	84,522	20.61	83.12	130.03
Safety duplication	2,230	120	3,805	7,266	0.05	1.71	3.26
Dissemination	3,680	1,019	26,700	23,554	0.28	7.26	6.40

Notes: Management costs (general management and information management) are allocated according to the following percentages for the CIMMYT genebank: medium-term storage (15%), long-term storage (15%), acquisition (5%), germination testing (10%), regeneration (35%), safety duplication (5%) and dissemination (25%).

[a]Total quasi-fixed costs include the costs of senior scientific and technical staff.
[b]Full capacities for wheat and maize are 390,000 and 67,000 accessions, respectively.

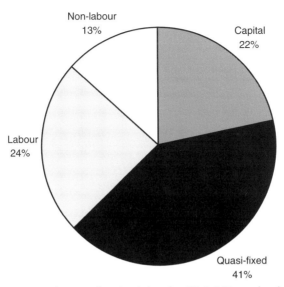

Fig. 3.1. Representative annual costs of maintaining the CIMMYT genebank holdings. Share of conservation and distribution costs by cost category (US$543,089, 1996 prices).

28% for wheat. These differences are largely attributable to the substantially higher amount of labour required to regenerate maize while limiting genetic drift, given its heterogeneous, outcrossing nature.

Economic costs

Annual average costs

Given that the genebank is operating well below capacity, the average costs per accession detailed in Table 3.7 provide upper-bound estimates of the corresponding marginal costs. It is these marginal costs that are central to assessing the economics of changes to the genebank operations on the margin or over the short run. For example, what is the cost of storing an existing accession for 1 more year, or, equivalently, what is the benefit in terms of cost savings from eliminating a duplicate accession from the genebank? The answer depends on the crop in question and on the state of the sample, including its time in storage, time from last regeneration or germination test, and such. If the sample is known to be viable, it costs little to hold over an accession of either crop for 1 more year – just 19 cents for an accession of wheat or 93 cents for an accession of maize. However, if the viability of the seed needs to be checked and then the sample regenerated because it failed the test, the cost of keeping it for another year jumps dramatically to $3.45 and $109.64 for each wheat and maize accession, respectively. Clearly, eliminating duplicate accessions could produce substantial cost savings. In fact, it would be economic to spend up to $109 to ascertain if a maize accession was duplicated in the CIMMYT holdings before regeneration (or, for that matter, held in collections at other sites, given the cost of shipping in seed, if needed, is comparatively low and falling).

Table 3.7. Average costs (US$ per accession, 1996 prices) of conserving and distributing an accession for 1 year at the CIMMYT genebank.

	Existing accession		New accession	
Cost category	Without regeneration	With regeneration	Without regeneration	With regeneration
Conservation				
Wheat	**0.19**	**3.45**	**3.61**	**8.08**
Long-term storage	0.19	0.19	0.19	0.19
New introduction				
Acquisition			2.99	2.99
Initial germination testing				1.21
Initial duplication			0.43	0.43
Germination testing		0.61		0.61
Regeneration		2.65		2.65
Maize	**0.93**	**109.64**	**13.88**	**126.84**
Long-term storage	0.93	0.93	0.93	0.93
New introduction				
Acquisition			7.99	7.99
Initial germination testing				4.25
Initial duplication			4.96	4.96
Germination testing		2.13		2.13
Regeneration		106.58		106.58
Distribution				
Wheat	**4.17**	**7.43**		
Medium-term storage	0.20	0.20		
Germination testing		0.61		
Dissemination	3.97	3.97		
Regeneration		2.65		
Maize	**15.28**	**123.99**		
Medium-term storage	1.62	1.62		
Germination testing		2.13		
Dissemination	13.66	13.66		
Regeneration		106.58		

A second policy question relates to the first: what is the cost of introducing a new accession into the genebank, given that the decision to store it is revisited after 1 year? The answer also depends on the protocol for new introductions of a specific crop as well as for the size and state of the sample. CIMMYT's standard procedure is to check the health status of virtually all incoming accessions. The sample size should be sufficient to enable storage in the genebank and provide a backup sample off site. If a new accession is viable and of sufficient size to negate the need to bulk up the sample, the cost to CIMMYT of incorporating this new accession in its genebank and storing it for 1 year is $3.61 per accession for wheat and $13.88 for maize. However, if the sample requires regenerating at the time of its introduction, this cost increases to $8.08 per accession for wheat and soars to $126.84 for maize. The corresponding costs for distribution are $4.17 for wheat and $15.28 for maize if regeneration is not required, and $7.43 for wheat and $123.99 for maize when low seed stock necessitates regeneration.

Another question relates to the comparative costs of keeping an accession for another year versus discarding it if the same accession could be reintroduced later from elsewhere if needed. According to our estimates, keeping the existing accession of wheat for 1 more year is clearly cheaper, providing the existing accession needs no regeneration to rebuild stocks or increase viability. If the existing accession needs regenerating because, for example, the sample size is too small, then the story is not clear-cut. For wheat it pays to roll over the existing accession for another year even if the introduced accession requires regenerating (as is usually the case with the small samples that are shipped among genebanks). For maize, however, while it is cheap to roll over an existing accession that needs no regeneration, the cost of introducing an accession from elsewhere versus keeping an accession that requires regeneration is comparable and substantial.

This type of cost calculus and its implied management responses are even more complex if we mimic a more realistic time–cost scenario. The dissemination data in Table 3.5 suggest that many genebank accessions sit untouched for many years. Indeed, it is the option value of these accessions rather than their more immediate use value that is the justification commonly cited for establishing and maintaining a genebank. However, that option value can only be realized if at some future date the sample is called upon for breeding or other research purposes. Rather than comparing the cost differentials of holding an existing accession versus introducing that same accession in the current year, a more subtle – but more relevant – question is the following: if an accession will be first utilized n years from now, how long must that delay, n, be before it is economical to rely on introductions from elsewhere rather than to maintain an existing holding?

Figures in Table 3.7 indicate that regeneration costs are high, especially for maize. If an existing maize accession requires regeneration and the same accession is known to be stored elsewhere, it may be more economical to discard the accession from the genebank unless it is utilized within 2–3 years (assuming that interest rates are in the 2–6% range). The cut-off period for wheat under the same situation is 7–10 years. Since the costs of introducing a wheat accession to the collection are large compared with the costs of storage, it is more economic to conserve existing accessions deemed useful in the near future. In the case where regeneration is not required, the cut-off periods are 7–10 years for wheat and 5–6 years for maize. In general, if accessions are unlikely to be used within a decade or so, it is better to store those accessions in a single facility and distribute them to local genebanks when requested, assuming transportation costs and other quarantine barriers are not prohibitive.

Average costs in the long run
Most of the figures above refer to the costs of conserving an accession for 1 more year, with the notion that decisions taken now can be revisited the following year. However, genebanks may well want, or be required, to guar-

antee safe keeping of samples in perpetuity – for example, those accessions held in trust by the CGIAR centres by way of their commitments to the FAO. The cost of such a guarantee obviously depends on the state of future technology, input costs (including the rate of interest), storage capacity and regeneration intervals.

Table 3.8 shows the average costs of conserving wheat and maize accessions in perpetuity, assuming that costs are constant over time in real (inflation-adjusted) terms. We considered the present values of the costs of conserving an existing accession and a newly introduced accession with different regeneration intervals and different real rates of interest (2, 4 and 6% per annum, which were deemed to span the relevant range). Testing for the viability of seed samples was assumed to begin 10 years after the introduction of a new accession to the collection, with retesting every 5 years thereafter.

The average cost of conserving an existing accession of wheat in perpetuity ranges from $6.33 to $22.66 when no initial regeneration occurs, and from $9.58 to $25.91 with an initial round of regeneration.[3] The present values of conservation are more sensitive to changes in the rate of interest than they are to changes in initial regeneration protocols: lower rates of interest result in higher present values of these cost streams; but the interest cost of securing this long-term commitment falls proportionally when interest rates are low. For maize, the comparable costs range from $32.28 to $151.46 per accession in the absence of an initial regeneration, and from $140.98 to $260.16 with an initial round of regeneration. As already established, regeneration constitutes a significantly larger share of overall conservation costs for maize than for wheat, so correspondingly larger cost consequences result from changes in the initial regeneration protocol in maize, especially at higher rates of interest. For example, at an interest rate of 6%, conserving a sample of maize seed in perpetuity costs $32.28 without an initial round of regeneration and $140.98 with initial regeneration, representing more than a 300% increase, whereas an accession of wheat costs $6.33 and $9.58, respectively, representing only about a 65% increase.

Above we considered costs at either end of the conservation spectrum – the marginal costs of conserving an accession for 1 more year versus the average costs of conserving an accession in perpetuity. Each type of cost is useful for different conservation and investment decisions. Marginal costs inform short-run (year-on-year) decisions to continue storing seeds, whereas the present values of average long-run (in-perpetuity) cost guide commitments to conserving the collection into the indefinite future. There are other possibilities involving the intermediate run. For example, one option is to keep the collection for the life of the genebank (taken here to be 40 years) and then consider alternatives such as abandoning the holding, rebuilding and perhaps extending the facility or shipping the seeds for storage elsewhere. This is a weaker commitment, so the present values of the corresponding costs are generally lower than those incurred in storing the collection in perpetuity. Table 3.9 is identical in format to Table 3.8 but provides cost data tailored to a 40-year conservation horizon. Once again, the

Table 3.8. Present values (US$ per accession, 1996 prices) of conserving and distributing an accession in perpetuity at the CIMMYT genebank.

Cost category	Without initial regeneration			With initial regeneration		
	2%	4%	6%	2%	4%	6%
Conservation						
Wheat						
Long-term storage	9.94	5.07	3.44	9.94	5.07	3.44
New introduction						
Acquisition	2.99	2.99	2.99	2.99	2.99	2.99
Initial germination testing	1.21	1.21	1.21	1.21	1.21	1.21
Initial duplication	0.43	0.43	0.43	0.43	0.43	0.43
Germination testing	10.54	4.60	2.68	10.54	4.60	2.68
Safety duplication	0.25	0.07	0.02	0.25	0.07	0.02
Regeneration[a] (50 years)	1.93	0.53	0.19	5.18	3.79	3.44
Conservation cost						
Existing accession	**22.66**	**10.27**	**6.33**	**25.91**	**13.53**	**9.58**
New accession	**27.29**	**14.90**	**10.96**	**30.54**	**18.16**	**14.21**
Maize						
Long-term storage	47.25	24.09	16.37	47.25	24.09	16.37
New introduction						
Acquisition	7.99	7.99	7.99	7.99	7.99	7.99
Initial germination testing	4.25	4.25	4.25	4.25	4.25	4.25
Initial duplication	4.96	4.96	4.96	4.96	4.96	4.96
Germination testing	37.03	16.14	9.40	37.03	16.14	9.40
Safety duplication	2.93	0.81	0.28	2.93	0.81	0.28
Regeneration (50 years)	64.26	17.80	6.24	172.96	126.50	114.94
Conservation cost						
Existing accession	**151.46**	**58.83**	**32.28**	**260.16**	**167.53**	**140.98**
New accession	**168.67**	**76.04**	**49.49**	**277.37**	**184.74**	**158.19**
Distribution						
Wheat						
Medium-term storage				9.98	5.09	3.46
Germination testing				10.54	4.60	2.68
Dissemination[b]				22.08	12.23	8.98
Regeneration (25 years)[c]				8.34	5.21	4.25
Distribution cost				**50.94**	**27.13**	**19.37**
Maize						
Medium-term storage				82.42	42.02	28.55
Germination testing				37.03	16.14	9.40
Dissemination				76.01	42.09	30.92
Regeneration (25 years)				278.39	173.96	141.72
Distribution cost				**473.85**	**274.21**	**210.59**

Note: Some figures in this table were calculated using equations (1), (2) and (4) in Appendix A.
[a]Regeneration costs include germination testing after every regeneration.
[b]Dissemination occurs every 10 years, based on the 5-year average in Table 3.5.
[c]Regeneration for dissemination commences in the 25th year, then occurs every 50 years.

Table 3.9. Present values (US$ per accession, 1996 prices) of conserving and distributing an accession for the life of the CIMMYT genebank.

Cost category	Without initial regeneration			With initial regeneration		
	2%	4%	6%	2%	4%	6%
Conservation						
Wheat						
Long-term storage	5.44	4.01	3.11	5.44	4.01	3.11
New introduction						
Acquisition	2.99	2.99	2.99	2.99	2.99	2.99
Initial germination testing	1.21	1.21	1.21	1.21	1.21	1.21
Initial duplication	0.43	0.43	0.43	0.43	0.43	0.43
Germination testing	4.72	3.18	2.21	4.72	3.18	2.21
Regeneration (50 years)	0.00	0.00	0.00	3.26	3.26	3.26
Conservation cost						
Existing accession	**10.16**	**7.19**	**5.32**	**13.42**	**10.45**	**8.58**
New accession	**14.79**	**11.82**	**9.95**	**18.05**	**15.08**	**13.21**
Maize						
Long-term storage	25.85	19.07	14.78	25.85	19.07	14.78
New introduction						
Acquisition	7.99	7.99	7.99	7.99	7.99	7.99
Initial germination testing	4.25	4.25	4.25	4.25	4.25	4.25
Initial duplication	4.96	4.96	4.96	4.96	4.96	4.96
Germination testing	16.59	11.17	7.76	16.59	11.17	7.76
Regeneration (50 years)	0.00	0.00	0.00	108.70	108.70	108.70
Conservation cost						
Existing accession	**42.44**	**30.24**	**22.54**	**151.14**	**138.94**	**131.24**
New accession	**59.64**	**47.44**	**39.74**	**168.34**	**156.14**	**148.44**
Distribution						
Wheat						
Medium-term storage				5.46	4.03	3.12
Germination testing				4.72	3.18	2.21
Dissemination				13.88	10.51	8.49
Regeneration (25 years)				5.24	4.48	4.02
Distribution cost				**29.30**	**22.20**	**17.84**
Maize						
Medium-term storage				45.10	33.27	25.78
Germination testing				16.59	11.17	7.76
Dissemination				47.77	36.17	29.24
Regeneration (25 years)				174.96	149.48	134.03
Distribution cost				**284.42**	**230.09**	**196.81**

Note: Some figures in this table were calculated using equations (1), (2) and (4) in Appendix A. The life of the genebank is assumed to be 40 years for the purposes of this study.

present value of variable and capital costs is sensitive to the rate of interest: the cost savings from committing to store seeds for 40 years instead of an infinite number of years contract as the rate of interest increases. For example, comparing the corresponding estimates in Tables 3.8 and 3.9, the cost of conserving an accession of wheat for 40 years versus forever reduces by more than 50% at a 2% interest rate (from $22.66 to $10.16 per accession), but reduces by less than 15% at a 6% interest rate (from $6.33 to $5.32 per accession). As the interest rate increases, the costs of storage beyond 40 years have smaller weight in the total cost.

Total costs in the short and long run

Table 3.10 illustrates the total costs of conserving seeds over different time horizons. Annualized fixed costs (including a prorated overhead labour component) provide an estimate of the average costs of storing all accessions for 1 year. For wheat, the costs are $152,076; for maize, they are $193,704. The corresponding costs of distribution are $101,052 for wheat and $101,951 for maize.

A commitment to conserve the accessions for the longer run naturally carries a higher price tag, with costs including storing and maintaining the viability of the collection. In present-value terms, the total costs of conserv-

Table 3.10. Total costs (US$, 1996 prices) of conserving existing accessions in the short and long run at the CIMMYT genebank.

Cost category	Conservation (without initial regeneration)	Conservation (with initial regeneration)	Distribution	Total
Wheat (123,000 accessions)				
1 year	**34,486**	**152,076**	**101,052**	**253,128**
Variable	8,526	41,052	29,952	71,004
Quasi-fixed	6,087	65,440	49,501	114,941
Capital	19,873	45,584	21,599	67,183
40 years	**709,867**	**3,130,398**	**2,080,094**	**5,210,492**
Variable	175,495	845,028	616,541	1,461,569
Quasi-fixed	125,307	1,347,052	1,018,948	2,366,000
Capital	409,065	938,318	444,605	1,382,923
In perpetuity	**896,625**	**3,953,967**	**2,627,340**	**6,581,307**
Variable	221,666	1,067,345	778,745	1,846,090
Quasi-fixed	158,274	1,701,444	1,287,020	2,988,464
Capital	516,685	1,185,178	561,575	1,746,753
Maize (17,000 accessions)				
1 year	**35,482**	**193,704**	**101,951**	**295,655**
Variable	8,902	86,875	37,339	124,214
Quasi-fixed	6,848	72,289	41,158	113,447
Capital	19,732	34,540	23,454	57,994
40 years	**730,385**	**3,987,285**	**2,098,608**	**6,085,893**
Variable	183,251	1,788,271	768,613	2,556,884
Quasi-fixed	140,971	1,488,023	847,214	2,335,237
Capital	406,163	710,991	482,781	1,193,772
In perpetuity	**922,539**	**5,036,288**	**2,650,724**	**7,687,012**
Variable	231,462	2,258,742	970,825	3,229,567
Quasi-fixed	178,058	1,879,503	1,070,105	2,949,608
Capital	513,019	898,043	609,794	1,507,837

ing CIMMYT's present maize holdings for the life of the genebank (under baseline assumptions with initial regeneration) are $3,130,398 for the wheat collection and $3,987,285 for the maize collection – a total of $7,117,683. Committing to conserve the seeds in perpetuity costs $3,953,967 for the wheat collection and $5,036,288 for maize – a total of $8,990,255.

For management and various policy and investment purposes, it is useful to break out the capital costs from the other expenses. According to our estimates, the present-value equivalent of $1,649,309 is needed to underwrite the capital costs of conserving CIMMYT's current maize and wheat holdings for the life of the genebank – a total of $2,083,221 if the seeds are conserved in perpetuity and the genebank facility and other capital items are replaced on a recurring basis as needed. Setting aside the cost of capital, it takes $5,468,374 in total labour and operating costs (including the labour of senior scientific staff) to conserve the entire wheat and maize holdings for 40 years, and $6,907,034 if the seeds are saved in perpetuity. This figure includes much more than the labour and operational costs required to simply store the seeds in the genebank. It factors in the costs of checking viability, regenerating the samples (presumed here to be a 50-year cycle, although certain seeds would need regeneration more frequently), plus the data-management costs required to manage the collection.

Separate from these costs are the costs of distributing the seeds. If the genebank continued to distribute seed at the rate typical of the past few years, this function alone would cost about $4,178,702 in present-value terms over a 40-year time horizon, and $5,278,064 in perpetuity. Bundling all these costs together (that is, including seed storage, regeneration and duplication; information management; and dissemination activities), we estimate that the capital, labour and operational costs combined would total $11,296,385 over the life of the genebank and $14,268,319 in perpetuity. This sizeable, but not especially large, sum of money represents the amount that would need to be set aside, at a 4% real interest rate, to underwrite genebank activities at their current levels over the longer run.

Notes

[1]Beginning in mid-1996, CIMMYT staff gradually began transferring maize and wheat seeds into the new facility. During this process the maize collection was checked for consistency with the genebank records and repacked into new containers in readiness for storage. Approximately 40,000 wheat accessions obtained from or regenerated in Karnal bunt-free areas were directly moved to the new facility. The process of regenerating the remaining 80,000 wheat accessions began in 1996 and is expected to be completed by 2006.

[2]For example, in 1996 an exceptionally large number of wheat accessions (22,000) were regenerated in the process of moving material from the old to the new genebank facility, while ensuring that the introductions to the new facility were free of Karnal bunt. An exceptionally large number of wheat accessions (35,000 in total) were also duplicated that year for backup storage purposes. We took the average number of accessions processed or stored over the past few years as our representa-

tive accession totals. For rescaling the new introductions estimates, we used 5000 wheat accessions (and 1000 maize accessions), germination testing was 6000 wheat (4000 maize), regeneration was 6000 wheat (500 maize), dissemination was 13,500 wheat (3800 maize) and duplication was 11,600 wheat (1500 maize).

[3]The format of these and all other estimates cited in the text is not to be misconstrued as implying any false sense of precision. They are cited this way to facilitate cross-referencing with the appropriate table.

4 ICARDA Genebank

Bonwoo Koo, Philip G. Pardey, Jan Valkoun and
Brian D. Wright

History of the ICARDA Genebank

ICARDA was established in Aleppo, Syria, in 1977. The institute targets the entire developing world in its crop-improvement research on barley, lentil and faba bean, and dry-land areas in developing countries in research aimed at improving the management of rangelands and water resources and the nutrition and productivity of small ruminants (such as sheep and goats) within the CG system. It also takes responsibility for breeding, improving farming systems and protecting and enhancing the natural resource base of water, land and biodiversity in the region of Central and West Asia and North Africa (CWANA). Including the subtropical and temperate dry-land areas within developing countries, ICARDA collectively covers an estimated one-third of the world's agricultural land. Rich in genetic resources in the region, ICARDA has paid particular attention to conserving landraces and other primitive materials of crops originating in the region (Valkoun et al., 1995).[1]

In 1983, ICARDA's various genetic resources activities were consolidated in a new Genetic Resources Unit (GRU), which was established to oversee the centre's germplasm collection, conservation, documentation, characterization and distribution activities. In addition, the unit provides a centre-wide seed-health service through its seed-health laboratory. In 1989, with financial support from the Italian government, the GRU completed construction of its new facilities, which include cold rooms for medium- and long-term storage of seed samples. IBPGR (the forerunner of IPGRI, as mentioned in Chapter 1) designated ICARDA as the holder of global base collections of barley, wild wheat relatives, durum wheat, lentil and faba bean and a base collection of bread wheat and chickpea in a coordinated regional network.

Initially, GRU mainly supported ICARDA's crop-improvement pro-grammes, but gradually interactions with national agricultural research agencies in the CWANA region became equally important. This process cul-minated in 1992 when ICARDA established a network for collaboration in plant genetic resources in the CWANA region with IBPGR and the FAO Commission on Plant Genetic Resources. In 1996, ICARDA, IPGRI and five Central Asian countries of the former Soviet Union set up another sub-regional network to collaborate on the conservation of genetic resources (the Central Asia Network for Plant Genetic Resources, CAN/PGR). Three transcaucasian countries joined this network in 1999, and its name was changed to the Central Asia and Transcaucasian Network for Plant Genetic Resources (CATN/PGR).

ICARDA gives high priority to collecting and conserving indigenous germplasm, not only from the Near Eastern centres of diversity but also from other countries in the CWANA region. Local germplasm is well adapted to the harsh, stressful and highly variable environments and may provide useful genes for breeding stress-tolerant varieties adapted to vari-ous target areas and farming systems. In addition, given the rapid pace of genetic erosion throughout the region from the loss of natural habitats for wild species, it became important to build up the collection of wild relatives and landraces in an *ex situ* facility.

As of 1998, the ICARDA genebank housed almost 120,000 accessions in total, about 50% of which were cereals, with the rest divided almost equally between food legumes and forage legumes (see Table 4.1). The majority of the collection (about 70%) came from major genebanks outside CWANA or directly from national programmes throughout the region. The major sources of ICARDA's holdings are the US Department of Agriculture's National Small Grains Collection in Aberdeen, Idaho (23,800 accessions), the Germplasm Institute in Bari, Italy (16,500 accessions), and the N.I. Vavilov All-Russian Research Scientific Institute of Plant Genetic Resources. About 24,500 accessions stem from 150 germplasm collection missions in cooperation with national programmes in 28 countries, mostly in CWANA. This collection activity continues in areas of substantial genetic diversity as well as in 'undersampled' areas throughout CWANA.[2]

In recent years, GRU has conducted research to study the genetic diversity within its collections of barley, durum wheat, lentil, chickpea and medicago, using agronomic, biochemical and molecular characterization techniques (such as randomly amplified polymorphic DNA (RAPD) and amplified frag-ment length polymorphism (AFLP) markers). This research provides a better understanding of the geographical distribution of crop diversity and is also being used to identify a core collection that involves a reduced number of accessions to represent the global diversity in each respective crop.

Research on genetic diversity in natural populations of wild progeni-tors and relatives of mandate crops also receives high priority because of its role in helping to identify the most representative and appropriate popula-tions to conserve in their natural habitats (*in situ* conservation). Such popu-lations have already been identified for wild progenitors of cultivated

Table 4.1. ICARDA genebank holdings, 1989–1998.

Crop	No. of accessions									
	1989	1990	1991	1992	1993	1994	1995	1996	1997	1998
Cereals	**45,810**	**46,993**	**50,600**	**51,396**	**52,933**	**54,490**	**54,947**	**55,341**	**56,546**	**57,900**
Barley	18,244	18,467	20,943	20,952	21,400	22,514	22,622	22,678	22,909	23,163
Wild *Hordeum*	1,338	1,354	1,529	1,545	1,584	1,625	1,689	1,693	1,730	1,740
Durum wheat	16,477	16,683	17,115	17,319	17,707	17,772	17,776	17,848	17,888	18,943
Bread wheat	6,897	7,272	7,569	7,613	7,737	7,974	8,054	8,277	8,959	8,962
Other cultivated *Triticum*	258	269	295	458	472	473	473	473	481	481
Wild *Triticum*	1,029	1,046	1,075	1,171	1,238	1,281	1,425	1,433	1,448	1,448
Aegilops	1,567	1,902	2,074	2,338	2,795	2,851	2,908	2,939	3,131	3,163
Food legumes	**24,177**	**24,746**	**25,386**	**26,395**	**26,492**	**27,282**	**27,926**	**28,378**	**29,301**	**29,905**
Chickpea	8,028	8,182	8,603	8,806	8,809	9,541	9,703	9,707	10,061	10,117
Wild *Cicer*	233	240	243	249	254	261	262	262	262	262
Lentil	6,804	7,164	7,164	7,277	7,315	7,337	7,598	7,727	7,753	8,261
Wild *Lens*	313	324	366	385	400	412	459	459	462	462
Faba bean	3,552	3,589	3,763	4,422	4,448	4,463	4,636	4,955	5,495	5,535
Faba beans – BPL	5,247	5,247	5,247	5,256	5,266	5,268	5,268	5,268	5,268	5,268
Forage legumes	**19,831**	**21,084**	**22,817**	**24,897**	**25,499**	**27,671**	**29,710**	**29,811**	**31,398**	**31,717**
Medicago annual	4,931	5,268	5,808	7,231	7,321	7,614	7,700	7,712	7,748	7,748
Vicia	4,158	4,267	4,461	4,565	4,901	5,141	5,423	5,462	5,558	5,567
Lathyrus	1,256	1,303	1,348	1,416	1,527	1,601	2,921	2,939	3,043	3,043
Trifolium	2,500	2,873	3,145	3,212	3,215	4,344	4,363	4,363	4,589	4,589
Pisum	3,539	3,612	3,665	3,800	3,850	3,868	4,023	4,051	4,680	4,990
Other forages	3,447	3,761	4,390	4,673	4,685	5,103	5,280	5,284	5,780	5,780
Total	**89,818**	**92,823**	**98,803**	**102,688**	**104,924**	**109,443**	**112,583**	**113,530**	**117,245**	**119,522**

BPL, Breeders' Primary Lines.

wheat, *Triticum urartu* and *Triticum dicoccoides,* and for wild lentil. To learn more about the factors important for *in situ* conservation, experiments have also begun in which self-regenerating populations of the wild ancestors of wheat, barley, lentil, chickpea and medicago were established in two different environments: long-term fallow and rangeland. Similarly, larger-scale experiments were established in collaboration with the Syrian Agricultural Research Centre at its three research stations.

ICARDA, IPGRI and the national programmes of Jordan, Lebanon, the Palestinian Authority and Syria developed a project to promote *in situ* and on-farm conservation of wild crop relatives and landraces in the Near Eastern centre of crop origin, also the primary location of diversity. The project, which involves substantial national and community participation, is funded by the Global Environmental Facility (GEF) and was implemented in 1999 with ICARDA as the executing and coordinating agency.

Costing the ICARDA Genebank

Capital input costs

Table 4.2 provides a breakdown of the capital input costs incurred by GRU. The last column provides annualized user costs for these capital items in constant dollars, using a 4% real interest rate along with replacement costs and service life estimates. The medium-term storage room is equipped with a refrigeration system (dehumidifiers, cooling devices and control unit) and various shelves and trays. The long-term storage room is equipped with a similar system without dehumidifiers. Seeds for long-term storage are packed in vacuum-sealed containers, costing 70 cents each. Plastic containers, costing 8–26 cents each, are used for medium-term storage.[3]

The seed-health laboratory tests all incoming and outgoing accessions for seed-borne pathogens and pests to ensure a disease-free collection. Housed in a separate building, the laboratory is equipped with essential equipment for laboratory testing and two greenhouses for field testing and quarantine purposes. Part of the laboratory's operation supports the centre's breeding programme, so the costs of those activities are excluded here. GRU also has two greenhouses and 20 screenhouses (called cages) for the regeneration of seeds requiring special attention. The general capital items in Table 4.2 include inputs (such as office space, computers and other office equipment) that are commonly shared among crops or are difficult to allocate to a specific crop.

Annual operating costs

Seed storage
The seeds in the medium-term storage facility (the active collection) are dried to 6–7% moisture content and maintained at 0–4°C with a relative humidity of 15–20%. Seeds in long-term storage (the base collection) are also dried to 6–7% moisture content and stored in a room with a

Table 4.2. Capital input costs (US$, 1998 prices) at the ICARDA genebank.

Cost category	Service life (years)	Replacement cost	Annualized cost[a]
Medium-term storage		279,186	24,352
Storage facility	40	104,500	5,077
Storage equipment	10	155,250	18,405
Seed containers	50	19,436	870
Long-term storage		317,479	22,797
Storage facility	40	68,600	3,333
Storage equipment	10	107,800	12,780
Vacuum-sealing device	10	5,000	593
Seed containers	50	136,079	6,091
Viability testing		40,000	3,133
Viability-testing facility	40	23,000	1,117
Incubator	10	12,000	1,423
Other equipment	10	5,000	593
Regeneration		239,904	28,043
Farming equipment	10	1,000	119
Greenhouse/screenhouse	10	112,554	13,343
Seed-cleaning equipment	10	2,150	255
Seed-drying facility	40	5,700	277
Seed-drying equipment	10	16,000	1,897
Seed-processing facility	40	37,500	1,822
Seed-processing equipment	10	2,000	237
Vehicles	7	63,000	10,093
Seed-health testing		57,000	5,314
Seed-health facility	40	25,500	1,239
Greenhouse	10	13,600	1,612
Lab/office equipment	10	12,000	1,423
Computer	5	1,700	367
Vehicle	7	4,200	673
General capital		192,500	19,737
General facility	40	113,000	5,490
Office equipment	10	30,000	3,556
Computers	5	49,500	10,691
Total capital cost		1,126,069	103,376

Note: See Appendix B for further details.
[a]Calculated at a 4% interest rate using equation (3) in Appendix A.

temperature of $-20°C$. Extending the viability of seeds held in a base collection using appropriate temperature and humidity controls reduces the frequency of regeneration, thus reducing the genetic drift that occurs during the regeneration process.

Non-labour storage costs in Table 4.3 include the costs of the electricity required to run the refrigeration system. Part of the relevant labour cost includes the time required by technicians to maintain and operate the storage facility and its equipment, but a major labour cost component is the genebank manager's time, of which about 20% is spent on storage-related activities, with medium-term storage requiring more time than long-term storage.

Table 4.3. Annual operating costs (US$, 1998 prices) of conservation and distribution at the ICARDA genebank.

Cost category	Labour	Non-labour	Subtotal	Capital
Acquisition	**3,977**	**456**	**4,433**	**402**
Seed-health testing	1,198	367	1,565	–
Seed handling	2,007	–	2,007	–
Overheads	772	89	861	–
(Number of accessions)			*(2,270)*	
Medium-term storage	**10,716**	**2,549**	**13,266**	**24,351**
Storage management	5,933	–	5,933	–
Climate control	2,702	2,054	4,757	–
Overheads	2,081	495	2,576	–
(Number of accessions)			*(119,520)*	
Long-term storage	**4,691**	**1,699**	**6,390**	**22,796**
Storage management	1,978	–	1,978	–
Climate control	1,802	1,369	3,171	–
Overheads	911	330	1,241	–
(Number of accessions)			*(83,200)*	
Viability testing	**5,524**	**248**	**5,772**	**3,133**
Viability testing	4,451	200	4,651	–
Overheads	1,073	48	1,121	–
(Number of accessions)			*(6,700)*	
Dissemination	**33,291**	**9,692**	**42,982**	**4,911**
Dissemination management	11,865	–	11,865	–
Seed-health testing	14,617	4,483	19,099	–
Packing/shipping	344	3,327	3,671	–
Overheads	6,465	1,882	8,347	–
(Number of accessions)			*(27,700)*	
Duplication	**2,583**	**9,234**	**11,817**	–
Packing/shipping	2,081	7,441	9,522	–
Overheads	502	1,793	2,295	–
(Number of accessions)			*(8,410)*	
Information management	**54,554**	**6,950**	**61,504**	–
Database management	43,960	2,400	46,360	–
Other expenses	–	3,200	3,200	–
Overheads	10,594	1,350	11,944	–
General management	**77,029**	**11,838**	**88,867**	**19,737**
Managerial staff	62,070	–	62,070	–
Electricity	–	2,739	2,739	–
Other expenses	–	6,800	6,800	–
Overheads	14,959	2,299	17,258	–
Total operating cost	**192,365**	**42,666**	**235,031**	**75,330**

Note: See Appendix B for further details.

Seed-health testing

Bringing new germplasm into the collection runs the risk of introducing exotic or otherwise undesirable pathogens. Similarly, disseminating germplasm can inadvertently spread pathogens to areas where they were not previously a problem. To avoid the exchange of infected accessions, the seed-health laboratory tests all genetic materials entering ICARDA or disseminated elsewhere for seed-borne pathogens and pests (El-Ahmed, 1998). After fumigation or cold treatment, all incoming seeds are visually inspected and tested in the laboratory. All seeds deemed to be free of pathogens are planted in post-quarantine areas for one generation.

For accessions that are to be disseminated, random seed samples are tested. Particular attention is paid to samples from fields where potential seed-borne diseases were found during field inspection. If a recipient country has specified certain pathogens or pests as quarantine organisms, appropriate tests are conducted to ensure disease-free seeds. After all these testing procedures, the seed-health laboratory applies for a phytosanitary certificate (issued by the Syrian government) and prepares a certificate of origin that accompanies the shipped seeds.

The operations of the seed-health laboratory are supervised by a (part-time) plant pathologist, who is assisted by a research associate, a technician and five temporary labourers hired on a daily basis. About 20% of the seed-health laboratory's activities are geared to genebank operations. Since the costs for testing the health of incoming and outgoing accessions are about the same, these costs were allocated to the acquisition and dissemination categories in Table 4.3 based on the respective shares of acquired and disseminated accessions in the genebank.

Viability testing

As with all genebanks, the viability of stored seeds needs to be checked periodically. About 70% of ICARDA's holdings of those crops for which it has a global mandate within the CG (barley, lentil and faba bean) and a regional CG mandate (chickpea and wheat) have had at least one round of viability testing. For crops of lower priority (e.g. some forage legumes), the proportion of accessions that have been tested for viability is quite low. Over the past 10 years, a total of 67,405 accessions have gone through viability testing, averaging 6700 accessions per year.

To test for viability, 40 seeds from each accession are placed in four separate Petri dishes, held in an incubator for a week and then visually checked. A large share of the testing costs consists of the labour required to supervise and carry out the tests. If the viability rate of an accession falls below 85% it is regenerated in the next season. The interval of time for subsequent rounds of viability testing has not been formulated, but a testing protocol is under development.

Regeneration

One of the most labour-intensive aspects of conserving germplasm is regeneration of stored accessions to maintain adequate viable stocks. Multiplication to scale up accessions is done when the holding falls below a minimum sample size (500–700 seeds, depending on the crop), and, as mentioned, regeneration is performed when seed viability drops below 85%. In addition, some accessions are planted out for evaluation and characterization purposes.[4] Other accessions such as wild wheat were multiplied for prebreeding purposes in 1998, the base year for our data. Newly acquired accessions are also planted in the field or greenhouse to increase the number of seeds per sample before placing them in storage.[5]

Regeneration activities are supervised by two germplasm specialists for each group of crops (that is, cereals, food legumes and forage legumes), and together they process around 4000–6000 accessions per year for each group. Most cultivated cereals and food legumes (lentil and chickpea) meet the recommended storage quantity after the first cycle of regeneration; however, some crops, such as wild cereal, faba bean and forage legumes, often need more than one cycle. We assumed two rounds of regeneration were required for these crops to guarantee an appropriate sample size, and adjusted the number of accessions accordingly.[6]

Tables 4.4 and 4.5 detail the costs incurred in the field operations and the seed-processing activities during the regeneration process for each crop. The cost of field management includes 80% of two crop specialists' labour; we allocated the remaining 20% to characterization activities. The steps undertaken prior to planting include land preparation, dressing seeds with fungicides and scarifying the hard cover from seeds of various legumes (such as some *Vicia* and other pasture species) to enable germination. The station operation unit maintains ICARDA's farmlands and provides land-preparation services for planting on a $400/ha charge-back basis. A scarification machine is used to scarify some species with large-sized seeds, and a more labour-intensive hand-scarification technique is required for small-sized seeds of wild food legumes and pastures.

Most accessions are planted in field plots of 3–4 m^2 in size, which contain two rows of planted seeds and two to three empty rows to prevent the mixing of seeds between adjacent plots.[7] Some crops require more specialized care and are planted in either plastic-covered greenhouses or meshed screenhouses. For example, *Medicago* and other pasture species with a small number of seed samples (fewer than 50 seeds) are planted out in greenhouses, while faba bean are planted in screenhouses to control cross-pollination. Hand-planting is preferred in most cases to avoid inadvertently mixing seeds, and some small-sized seeds are not suitable for machine planting.

The amount of chemicals (fertilizers, herbicides and pesticides) applied each year depends on the incidence of disease infestations, which in turn depends on temperature and other environmental factors. Additionally, hand-weeding is also used when necessary because herbicides are effective only for some specific types of weeds (for example, broad-leaf or narrow-leaf weeds). Plants are inspected throughout the growing period to characterize their morphological traits. This activity requires a total of about 2 months of a specialist's and 2 months of a daily labourer's time for each group of crops.

Harvesting is done with the help of the station operation unit, which provides a driver for harvester machines. Cultivated cereals and chickpea are harvested by machine, but lentil and faba bean are hand-harvested, as

Table 4.4. Annual costs (US$, 1998 prices) of regeneration and characterization at the ICARDA genebank: cereals and forage legumes.

Cost category	Cereals				Forage legumes			
	Labour	Non-labour	Subtotal	Capital	Labour	Non-labour	Subtotal	Capital
Regeneration	**23,253**	**953**	**24,206**	**4,955**	**27,015**	**6,235**	**33,250**	**16,255**
Field operation	**17,288**	**953**	**18,241**	**3,364**	**19,764**	**6,235**	**25,999**	**14,446**
Field management	12,320	–	12,320	3,364	12,320	–	12,320	3,364
Seed preparation	74	50	124	–	241	70	311	119
Land preparation	1,120	480	1,600	–	980	420	1,400	–
Planting	37	–	37	–	333	–	333	–
Screenhouse	–	–	–	–	1,090	578	1,668	10,963
Chemicals/weeding	296	142	438	–	370	2,372	2,742	–
Harvesting	84	96	180	–	592	1,584	2,176	–
Overheads	3,357	185	3,542	–	3,838	1,211	5,049	–
Seed processing[a]	**5,965**	–	**5,965**	**1,591**	**7,251**	–	**7,251**	**1,809**
Processing management	4,206	–	4,206	730	4,781	–	4,781	830
Seed cleaning	389	–	389	90	858	–	858	103
Seed drying	53	–	53	771	81	–	81	876
Medium-term packing	70	–	70	–	56	–	56	–
Long-term packing	89	–	89	–	67	–	67	–
Overheads	1,158	–	1,158	–	1,408	–	1,408	–
(Number of accessions)			*(4,800)*				*(7,200)*	
Characterization	**4,281**	–	**4,281**	–	**4,741**	–	**4,741**	–
Recording traits	3,450	–	3,450	–	3,820	–	3,820	–
Overheads	831	–	831	–	921	–	921	–
(Number of accessions)			*(4,800)*				*(7,200)*	

Note: See Appendix B for further details.
[a] Costs of seed processing are based on the adjusted number of regenerated samples.

Table 4.5. Annual costs (US$, 1998 prices) of regeneration and characterization at the ICARDA genebank: chickpea, lentil and faba bean.

Cost category	Chickpea				Lentil				Faba bean			
	Labour	Non-labour	Subtotal	Capital	Labour	Non-labour	Subtotal	Capital	Labour	Non-labour	Subtotal	Capital
Regeneration	**5,109**	**1,138**	**6,247**	**1,456**	**5,218**	**1,237**	**6,455**	**3,358**	**12,437**	**2,761**	**15,198**	**2,019**
Field operation	**3,909**	**1,138**	**5,047**	**1,121**	**4,127**	**1,237**	**5,364**	**3,056**	**10,501**	**2,761**	**13,262**	**1,566**
Field management	2,710	167	2,877	1,121	2,710	167	2,877	1,121	6,899	167	7,066	1,121
Seed preparation	19	10	29	–	33	10	43	–	37	25	62	–
Land preparation	280	120	400	–	280	120	400	–	700	300	1,000	–
Planting	17	–	17	–	17	–	17	–	74	–	74	–
Screenhouse	–	–	–	–	192	102	294	1,935	641	320	961	445
Chemicals/weeding	37	400	437	–	37	400	437	–	–	1,017	1,017	–
Harvesting	87	220	307	–	56	198	254	–	111	396	507	–
Overheads	759	221	980	–	802	240	1,042	–	2,039	536	2,575	–
Seed processing[a]	**1,200**	–	**1,200**	**335**	**1,091**	–	**1,091**	**301**	**1,936**	–	**1,936**	**452**
Processing management	885	–	885	154	797	–	797	138	1,195	–	1,195	207
Seed cleaning	37	–	37	19	37	–	37	17	296	–	296	26
Seed drying	11	–	11	162	11	–	11	146	20	–	20	219
Medium-term packing	15	–	15	–	15	–	15	–	19	–	19	–
Long-term packing	19	–	19	–	19	–	19	–	30	–	30	–
Overheads	233	–	233	–	212	–	212	–	376	–	376	–
(Number of accessions)			*(1,000)*				*(900)*				*(1,800)*	
Characterization	**979**	–	**979**	–	**979**	–	**979**	–	**2,508**	–	**2,508**	–
Recording traits	789	–	789	–	789	–	789	–	2,021	–	2,021	–
Overheads	190	–	190	–	190	–	190	–	487	–	487	–
(Number of accessions)			*(1,000)*				*(900)*				*(1,800)*	

Note: See Appendix B for further details.
[a]Costs of seed processing are based on the adjusted number of regenerated samples.

are some wild forage legumes, because of seed shattering problems. Harvested materials are packed in paper bags (for cereal) or cotton bags (for legumes) and transported to the seed cleaning area. Seeds are then threshed and cleaned either manually (some wild forage species and pastures) or mechanically (cereals and some food legumes).

Upon receipt from the seed cleaning area, seed samples are first weighed, separated, and then placed in a dehumidifying room for 4–5 weeks. The room is held at 10–13% relative humidity and 21°C – designed to reduce the seed moisture content to a 6–7% range. After the seeds have been cleaned and dried, they are stored in plastic containers for medium-term storage and in aluminium bags for long-term storage.[8]

Dissemination and safety duplication

Growing interaction among national programmes in CWANA, scientists at ICARDA and partners in advanced institutions and other genebanks has increased requests for germplasm. Around 30,000 accessions have been disseminated to collaborators each year for the past 5 years, with usually 40–50% of these accessions being distributed to scientists outside ICARDA. The initial lines distributed to national programmes for testing were a selection based on trials at ICARDA's primary evaluation sites in Syria and Lebanon. In 1998, a total of 132 shipments involving 27,700 accessions were made, including 54 overseas shipments which require the phytosanitary certificate described above (Table 4.6). Shipping costs depend on the locations of recipients and the amount shipped. The genebank normally incurs between $3000 and $4000 per year in shipping costs.

ICARDA backs up its collection at various institutions. Duplicate samples of barley and wheat are sent to CIMMYT. The chickpea collection is duplicated at ICRISAT, the faba bean collection at the Austrian national genebank and the lentil collection at the National Bureau of Plant Genetic Resources (NBPGR) in New Delhi, India. *Medicago* and *Vicia* are conserved at the Federal Institute of Agrobiology (FIA) in Austria, and *Lathyrus* is duplicated at the Station Fédérale de Recherches Agronomiques de Changins (RAC) in Switzerland.

Duplication activities are overseen by the genebank manager. Not all accessions are duplicated every year. For example, 2097 food legumes and 6313 forage legumes were sent as duplicates to the above-mentioned locations in 1998. When the number of unique accessions requiring duplication reaches a certain level, seed samples are sent via the black-box procedure described in Chapter 3.

Information and general management

The documentation officer coordinates data gathered from the field and other sources, manages the information and oversees publication of occasional catalogues describing the holdings of each crop. When new accessions are introduced into the genebank, passport data for each collection are entered in the GRU information system as well as into ICARDA's Data

Table 4.6. Dissemination of germplasm from the ICARDA genebank, 1989–1998.

Crop/activity	1989	1990	1991	1992	1993	1994	1995	1996	1997	1998
Cereals										
Dissemination	**6,109**	**6,764**	**15,867**	**14,958**	**9,601**	**12,667**	**10,093**	**13,454**	**10,285**	**8,893**
GRU in ICARDA	1,441	1,023	3,191	1,532	1,285	2,653	4,114	3,413	3,609	191
Elsewhere in ICARDA	3,695	2,766	1,475	8,273	2,346	3,988	3,902	5,107	3,948	1,803
Outside ICARDA	973	2,975	11,201	5,153	5,970	6,026	2,077	4,934	2,728	6,899
Safety duplication	–	**6,134**	**8,815**	**655**	–	–	–	–	**8,068**	–
Food legumes										
Dissemination	**2,674**	**4,561**	**5,520**	**10,143**	**7,825**	**10,826**	**11,488**	**11,031**	**11,986**	**11,103**
GRU in ICARDA	1,161	1,744	2,705	4,998	2,400	4,054	4,575	4,434	4,304	2,763
Elsewhere in ICARDA	1,471	2,152	1,786	2,742	2,299	551	2,361	2,989	4,425	2,213
Outside ICARDA	42	665	1,029	2,403	3,126	6,221	4,552	3,608	3,257	6,127
Safety duplication	–	–	**595**	–	**1,000**	**4,000**	**6,676**	**1,000**	–	**2,097**
Forage legumes										
Dissemination	**370**	**1,341**	**2,325**	**5,233**	**7,027**	**8,937**	**8,028**	**9,344**	**8,894**	**7,715**
GRU in ICARDA	277	284	809	3,501	1,524	3,984	3,031	6,803	5,058	2,373
Elsewhere in ICARDA	43	717	616	441	3,109	1,679	3,171	513	2,079	1,071
Outside ICARDA	50	340	900	1,291	2,394	3,274	1,826	2,028	1,757	4,271
Safety duplication	–	–	–	–	–	–	–	**5,295**	–	**6,313**
Total dissemination	**9,153**	**12,666**	**23,712**	**30,334**	**24,453**	**32,430**	**29,609**	**33,829**	**31,165**	**27,711**
Total safety duplication	–	**6,134**	**9,410**	**655**	**1,000**	**4,000**	**6,676**	**6,295**	**8,068**	**8,410**

Management and Retrieval System. Passport, characterization and evaluation data are published regularly in germplasm catalogues.[9] The information is also available on the Internet as part of the SINGER project (see Chapter 3). An important part of GRU's data-management system is the seed-stock control system, which assists germplasm specialists in managing the storage of seeds and distribution of samples to users.

We determined that about 20% of genebank manager and administrative support-staff time was spent on activities other than the conservation and distribution efforts costed here. The remaining 80% was allocated to the general management category in Table 4.3. Various non-labour costs, including the costs of electricity and other miscellaneous operational expenses, were similarly allocated. However, the time spent by other senior staff (including genebank manager and germplasm specialists) was allocated directly to the crops and functions they oversee.

Economic Analysis

Representative annual costs of genebank operation

Table 4.7 provides an overview of the total capital, quasi-fixed and variable costs grouped by crop activities, along with corresponding average costs per accession (see 'The Simple Economics of Genebanking' in Chapter 2 for descriptive detail). The total annual cost of conserving and distributing germplasm at ICARDA is estimated to be about $437,247.

Figure 4.1 gives a cost component breakdown of this total. More than half of the annual expense of the genebank operation involves labour input, including quasi-fixed capital (63%), while less than a quarter of the total cost is related to capital (24%). Figure 4.1 also indicates that a substantial portion of the labour cost is lumpy, suggesting once again that, within a certain range, the overall costs of conservation do not increase dramatically as the number of accessions increases; hence the average cost per accession critically depends on the number of accessions.

Economic costs

Annual average costs
As previously established, determining the cost of storing an existing accession for 1 more year (or the savings from eliminating a duplicate accession from the genebank) depends primarily on the state of the sample and, in turn, on its time in storage (meaning the time from last regeneration or viability test). If a sample does not require regeneration, the average cost of holding over an accession of any crop for 1 more year at ICARDA is just 17 cents. If the sample does require regeneration, the average cost of keeping it for another year jumps to $5.29 for cereal and $8.64 for faba bean (Table 4.8).

Table 4.7. Annual total (US$, 1998 prices) and average (US$ per accession, 1998 prices) costs of each operation at the ICARDA genebank.

Cost category	Number of accessions	Total capital costs	Total quasi-fixed costs[a]	Total variable costs	Average capital costs	Average quasi-fixed costs	Average variable costs
Medium-term storage	119,520	27,312	24,674	11,147	0.23	0.21	0.09
Long-term storage	83,200	23,783	8,225	5,683	0.29	0.10	0.07
Acquisitions	2,270	1,389	9,459	2,493	0.61	4.17	1.10
Viability testing	6,700	4,120	10,679	2,611	0.61	1.59	0.39
Dissemination	27,700	10,832	64,409	23,686	0.39	2.33	0.86
Safety duplication	8,410	987	8,225	11,110	0.12	0.98	1.32
Regeneration							
Cereals	4,750	6,929	32,050	7,195	1.46	6.75	1.51
Forage legumes	5,400	18,228	32,764	15,522	3.38	6.07	2.87
Chickpea	1,000	2,114	8,310	2,949	2.11	8.31	2.95
Lentil	900	4,015	8,200	3,266	4.46	9.11	3.63
Faba bean	1,350	2,676	13,892	6,318	1.98	10.29	4.68
Characterization[b]							
Cereals	4,750	329	5,746	1,042	0.07	1.21	0.22
Forage legumes	5,400	329	5,746	1,501	0.06	1.06	0.28
Chickpea	1,000	110	1,482	332	0.11	1.48	0.33
Lentil	900	110	1,482	332	0.12	1.65	0.37
Faba bean	1,350	110	2,782	562	0.08	2.06	0.42
Total costs		103,373	238,125	95,749			

Notes: Management costs (general and information management) are allocated according to the following percentages for the ICARDA genebank: medium-term storage (15%), long-term storage (5%), acquisition (5%), viability testing (5%), regeneration (30%), characterization (5%), safety duplication (5%) and dissemination (30%).
[a]Total quasi-fixed costs include the cost of senior scientific and technical staff.
[b]Management costs for regeneration and characterization are allocated by the ratio of 8 : 2.

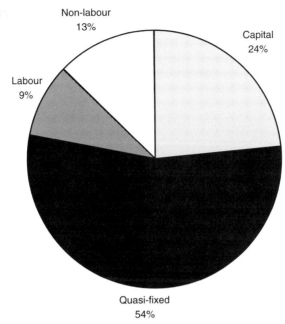

Fig. 4.1. Representative annual costs of maintaining the ICARDA genebank holdings. Share of conservation and distribution costs by cost category (US$437,247, 1998 prices).

Looking at the cost of conserving a newly acquired accession for 1 year, the average cost is between $14.84 and $18.19 per accession, including the regeneration costs incurred with checking seed health and stocking sufficient seed. Assuming sufficient stock, the average cost of distributing an accession is $3.48, which includes the costs of maintaining the active collection as well as packing and shipping. If the size of the holding is insufficient and regeneration is required, the average distribution cost jumps to between $8.60 and $11.96 per accession, depending on the crop. For a newly acquired accession, the one-time characterization cost increases the distribution cost to between $10.03 and $14.43 (Table 4.8).

Average costs in the long run
Table 4.9 provides present values for the costs of conserving and distributing an accession in perpetuity (in constant, inflation-adjusted terms as previously). The appropriate interval for viability testing has not yet been established for the current conservation technology; we assumed that testing begins in the tenth year after acquisition, with retesting every 5 years thereafter (consistent with CIMMYT's protocols). An average of around 25,000 accessions are disseminated each year, and so we assumed that an accession is disseminated once every 5 years. Using 4% as a baseline rate of interest, the average cost of conserving an existing accession of cereals in perpetuity is $10.14 per accession when no initial regeneration is needed. The comparable figures for other crops are in the range of $10.20–$10.69. For a newly acquired accession, the conservation costs range from $23.82 to $27.72 per accession.

Table 4.8. Average costs (US$ per accession, 1998 prices) of conserving and distributing an accession for 1 year at the ICARDA genebank.

Cost category	Existing accessions		New accessions (with regeneration)
	Without regeneration	With regeneration	
Conservation			
Long-term storage	0.17	0.17	0.17
New introduction			
Acquisition			5.27
Initial viability testing			1.98
Initial duplication			2.30
Viability testing		0.99	0.99
Regeneration[a]			
Cereals		4.13	4.13
Forage legumes		4.47	4.47
Chickpea		5.63	5.63
Lentil		6.37	6.37
Faba bean		7.49	7.49
Conservation cost by crop			
Cereals	**0.17**	**5.29**	**14.84**
Forage legumes	**0.17**	**5.63**	**15.18**
Chickpea	**0.17**	**6.79**	**16.34**
Lentil	**0.17**	**7.53**	**17.08**
Faba bean	**0.17**	**8.64**	**18.19**
Distribution			
Medium-term storage	0.30	0.30	0.30
Dissemination	3.18	3.18	3.18
Viability testing		0.99	0.99
Regeneration			
Cereals		4.13	4.13
Forage legumes		4.47	4.47
Chickpea		5.63	5.63
Lentil		6.37	6.37
Faba bean		7.49	7.49
Characterization			
Cereals			1.43
Forage legumes			1.34
Chickpea			1.81
Lentil			2.02
Faba bean			2.48
Distribution cost by crop			
Cereals	**3.48**	**8.60**	**10.03**
Forage legumes	**3.48**	**8.94**	**10.28**
Chickpea	**3.48**	**10.10**	**11.92**
Lentil	**3.48**	**10.84**	**12.86**
Faba bean	**3.48**	**11.96**	**14.43**

[a]Regeneration is done for both conservation and distribution concurrently, so the costs are equally allocated.

Table 4.9. Present values (US$ per accession, 1998 prices) of conserving and distributing an accession in perpetuity at the ICARDA genebank.

Cost category	Existing accession[a]			New accession		
	2%	4%	6%	2%	4%	6%
Conservation						
Long-term storage	8.53	4.35	2.95	8.53	4.35	2.95
New introduction						
Acquisition				5.27	5.27	5.27
Initial viability testing				0.99	0.99	0.99
Initial duplication				2.30	2.30	2.30
Viability testing[b]	9.53	4.58	2.93	9.53	4.58	2.93
Safety duplication[c]	1.36	0.38	0.13	1.36	0.38	0.13
Regeneration (50 years)						
Cereals	3.03	0.84	0.29	8.15	5.96	5.42
Forage legumes	3.23	0.89	0.31	8.69	6.36	5.78
Chickpea	3.91	1.08	0.38	10.54	7.71	7.00
Lentil	4.35	1.21	0.42	11.71	8.57	7.78
Faba bean	5.01	1.39	0.49	13.49	9.87	8.96
Conservation cost by crop						
Cereals	**22.44**	**10.14**	**6.31**	**36.12**	**23.82**	**19.99**
Forage legumes	**22.64**	**10.20**	**6.33**	**36.66**	**24.21**	**20.35**
Chickpea	**23.33**	**10.38**	**6.40**	**38.50**	**25.56**	**21.57**
Lentil	**23.77**	**10.51**	**6.44**	**39.68**	**26.42**	**22.36**
Faba bean	**24.42**	**10.69**	**6.50**	**41.46**	**27.72**	**23.54**
Distribution						
Medium-term storage	15.29	7.79	5.29	15.29	7.79	5.29
Dissemination[d]	33.74	17.86	12.58	33.74	17.86	12.58
Regeneration (25 years)						
Cereals	12.97	5.31	2.82	18.09	10.43	7.94
Forage legumes	13.83	5.66	3.01	19.29	11.13	8.47
Chickpea	16.76	6.87	3.64	23.38	13.49	10.26
Lentil	18.63	7.63	4.05	25.99	14.99	11.41
Faba bean	21.45	8.79	4.66	29.93	17.27	13.14
Characterization						
Cereals				1.43	1.43	1.43
Forage legumes				1.34	1.34	1.34
Chickpea				1.81	1.81	1.81
Lentil				2.02	2.02	2.02
Faba bean				2.48	2.48	2.48
Distribution cost by crop						
Cereals	**61.99**	**30.96**	**20.70**	**68.54**	**37.52**	**27.25**
Forage legumes	**62.85**	**31.32**	**20.88**	**69.65**	**38.12**	**27.69**
Chickpea	**65.78**	**32.52**	**21.52**	**74.21**	**40.95**	**29.96**
Lentil	**67.65**	**33.28**	**21.93**	**77.03**	**42.66**	**31.31**
Faba bean	**70.48**	**34.44**	**22.54**	**81.43**	**45.39**	**33.50**

Note: Some figures in this table were calculated using equations (1) and (2) in Appendix A.
[a]Existing accessions are assumed to have been freshly regenerated. New accessions require initial regeneration.
[b]Viability testing commences in the 10th year after acquisition, then every 5 years thereafter.
[c]Safety duplication is made concurrently with each round of regeneration.
[d]Dissemination occurs every 5 years.

Given that the active collection requires more frequent regeneration than the base collection (see 'Storing seeds and other plant material' in Chapter 2), we assume that accessions are regenerated every 25 years in medium-term storage and every 50 years in long-term storage. The present value of the distribution cost is in the range of $30.96–34.44 if regeneration to increase seed stock is not necessary. The added cost of characterizing newly acquired accessions increases the present value of distribution costs to between $37.52 and $45.39 per accession (Table 4.9).

Total costs in the long run

We reported in the previous section that the annual operating cost of the ICARDA genebank is $437,247 (Fig. 4.1). Using the in-perpetuity costs of conservation and distribution discussed in Table 4.9, Table 4.10 provides the total costs of conserving and distributing ICARDA's entire holdings in perpetuity. The table shows that it requires the present-value equivalent of $4,491,238 to conserve the present collection in perpetuity – $2,941,782 for non-capital such as labour and operating costs and $1,549,456 for capital. If the genebank continued to distribute seed at the rate typical of the past few years, this distribution function alone would incur an in-perpetuity cost of about $5,928,045 ($4,667,882 for non-capital and $1,260,163 for capital), expressed in present-value terms.

Bundling all these costs together (seed storage, regeneration, duplication, information management and dissemination activities), we estimate that the combined capital, labour and operational costs would total $10,419,283 in perpetuity ($7,609,664 for non-capital and $2,809,619 for capital), once again on the basis of setting this money aside at a 4% real rate of interest (Table 4.10).

Notes

[1]The CWANA region is particularly rich in genetic resources, and spans three of the eight centres-of-origin identified by Vavilov.

[2]Areas generally favoured for collection missions include those with a harsh climate, ecology, crop history and cultural practices that may endow the indigenous germplasm with useful traits for future crop-improvement research, such as tolerance to drought and extremes of temperature, resistance to pests and diseases, and various food or feed qualities.

[3]A total of 83,197 accessions (53,982 cereals, 16,727 food legumes and 12,488 forage legumes) were held in long-term storage in 1998 compared with 119,522 accessions in the medium-term storage facility.

[4]For example, the main focus of the cereal germplasm group for the past 3 years has been the preparation of a catalogue of its bread-wheat holdings, so substantial effort has been directed to characterizing and evaluating this crop.

[5]Less than 1% of incoming material does not require bulking up.

[6]In 1998, a total of 4800 accessions of cereal, 3700 accessions of food legumes and 7200 accessions of forage legumes were planted for regeneration and characterization. Some of these accessions represent carry-overs from the previous cycle of

Table 4.10. Total costs of conservation and distribution in perpetuity at the ICARDA genebank.

Crop	Number of accessions	Per-accession cost (US$ per accession, 1998 prices)			Total cost (US$, 1998 prices)		
		Conservation	Distribution	Total	Conservation	Distribution	Total
Cereals	57,900	**36.30**	**47.21**	**83.51**	**2,101,803**	**2,733,312**	**4,835,115**
Non-capital		23.82	37.52	61.34	1,379,077	2,172,132	3,551,209
Capital		12.48	9.69	22.17	722,726	561,180	1,283,906
Forage legumes	31,700	**37.81**	**49.76**	**87.56**	**1,198,626**	**1,577,288**	**2,775,914**
Non-capital		24.21	38.12	62.33	767,578	1,208,422	1,976,000
Capital		13.60	11.64	25.24	431,048	368,866	799,914
Chickpea	10,400	**38.42**	**51.35**	**89.77**	**399,628**	**534,065**	**933,693**
Non-capital		25.56	40.95	66.51	265,845	425,903	691,748
Capital		12.86	10.40	23.26	133,783	108,162	241,945
Lentil	8,700	**40.65**	**55.46**	**96.11**	**353,686**	**482,550**	**836,236**
Non-capital		26.42	42.66	69.08	229,888	371,163	601,051
Capital		14.23	12.80	27.03	123,798	111,387	235,185
Faba bean	10,800	**40.51**	**55.63**	**96.14**	**437,495**	**600,830**	**1,038,325**
Non-capital		27.72	45.39	73.11	299,394	490,262	789,656
Capital		12.79	10.24	23.03	138,101	110,568	248,669
All crops	**119,500**	**193.69**	**259.41**	**453.10**	**4,491,238**	**5,928,045**	**10,419,283**

regeneration. To allow for this fact, we estimated that the number of new accessions being regenerated in the current cycle were 4750, 3250 and 5400, respectively. It is these adjusted accession numbers that are used to scale up the per-accession cost to the corresponding total cost.

[7]One hectare can be divided into 1000 plots, so approximately 1000 accessions are planted per hectare.

[8]In most years, not all harvested accessions are cleaned, dried and stored: for example, some are used for testing and prebreeding purposes. However, we assumed that all of the multiplied accessions were processed and packed for storage when estimating the average costs of regeneration reported in Table 4.7.

[9]GRU also supports two international databases, one on wild wheat including *Aegilops*, and the other on forages from the Mediterranean basin, which was initially created through collaboration with IBPGR.

5   ICRISAT Genebank

BONWOO KOO, PHILIP G. PARDEY, N. KAMESWARA RAO AND
PAULA J. BRAMEL

History of the ICRISAT Genebank

During the 1960s, the Indian Agricultural Program of the Rockefeller
Foundation assembled over 16,000 sorghum germplasm accessions from
major sorghum-growing areas. ICRISAT acquired 8961 accessions of this
collection in 1974 through the All India Coordinated Sorghum
Improvement Project (AICSIP) based in Rajendranagar and another 3000
accessions from duplicate sets maintained in the United States (Purdue and
Fort Collins) and Puerto Rico (Mayaguez). It also acquired over 2000 pearl-
millet germplasm accessions assembled by the Rockefeller Foundation in
collaboration with the Indian Council of Agricultural Research (ICAR) in
New Delhi, and another 2000 accessions collected by the Institut de
Recherche pour le Développement (IRD, formerly the Institut Français de
Recherche Scientifique pour le Développement en Coopération (ORSTOM))
operations in francophone West Africa.

The chickpea and pigeonpea germplasm acquired by ICRISAT consists
of the materials originally collected and assembled by the former Regional
Pulse Improvement Project, a joint project of the Indian Agricultural
Research Institute (IARI), the US Department of Agriculture and Karaj
Agricultural University in Iran. Over 1200 chickpea accessions were
acquired from the Arid Lands Agricultural Development Program (ALAD)
with its headquarters in Beirut, supported by the Ford Foundation and the
International Development Research Centre in Canada. Similarly, much of
the groundnut germplasm initially assembled at ICRISAT was received
from the collections maintained by the Indian national programmes such as
the National Research Centre for Groundnut, as well as material held by the
US Department of Agriculture and North Carolina State University.

With over 113,500 accessions from 130 countries conserved in its
genebank, ICRISAT currently acts as a global repository for the genetic

© The International Plant Genetic Resources Institute (for the CGIAR System-wide Genetic Resources
Programme) and the International Food Policy Research Institute 2004. *Saving Seeds*
(B. Koo, P.G. Pardey and B.D. Wright *et al.*)

resources of its CG-mandated crops (sorghum, pearl millet, chickpea, pigeonpea and groundnut), their wild relatives and six minor varieties of millet (finger, foxtail, barnyard, kodo, little and proso millet). Between 1974 and 1997, ICRISAT launched 212 collection missions in areas of diversity, collecting a total of 8957 sorghum, 10,802 pearl millet, 4228 chickpea, 3870 pigeonpea and 2666 groundnut accessions. In addition to ICRISAT's own collection efforts and the material received from the major donors cited above, donations came from several other individuals, from national and international organizations and from the Ethiopian Sorghum Improvement Project, the Gezira Agricultural Research Station, AICSIP (Sudan), and several Indian agricultural universities (Table 5.1).

In addition to the main genebank at Patancheru in India, ICRISAT manages several regional genebanks in Africa that provide direct germplasm access to crop-improvement programmes in the region. Groundnut and pearl millet are stored at Niamey, Niger; sorghum and pearl millet at Bulawayo, Zimbabwe; and sorghum, pigeonpea and chickpea at Nairobi, Kenya. The Niamey genebank holds over 10,600 accessions, including about 1770 accessions of pearl millet, 5500 of groundnut, 1200 of groundnut breeding lines, 305 of sorghum and 1856 of forage grass. The base collection is held at 2°C and the active collection at 15°C and 15% relative humidity. The Bulawayo storage holds 12,000 sorghum and 6074 pearl-millet accessions in the active collection maintained at 4°C and 30% relative humidity.

Costing the ICRISAT Genebank

Capital input costs

The medium- and long-term storage facilities consist of several cooling modules, which are prefabricated panels equipped with cooling systems and dehumidifiers. These modules are housed in existing buildings. The medium-term storage consists of six modules and the long-term storage of three, including mobile and fixed shelving systems and centrally located electronic alarms and fire warning systems to monitor climate conditions and safeguard against fire hazard. The active collection is stored in variously sized, screw-topped aluminium cans or plastic jars; the base collection is stored in vacuum-sealed aluminium packets.

Viability testing is conducted in a seed laboratory, which is equipped with two seed germinators, two incubators and a water distiller. Farmland preparation and farming equipment for regeneration are operated and managed by the Farm and Engineering Service Program (FESP). The genebank manages its own greenhouse, which is used exclusively for the cultivation of wild species, especially wild groundnut. After harvesting, seeds are cleaned either by hand or mechanically with seed blowers and then dried in a seed drying room and packed.

Accessions being stored long-term or prepared for distribution are tested for seed health by the Plant Quarantine Unit (PQU), housed separately along with a fumigation building and 56-m² greenhouse. We allocated 50% of capi-

Table 5.1. ICRISAT genebank holdings, 1974–1999.

Year	Sorghum	Pearl millet	Chickpea	Pigeonpea	Groundnut[a]	Small millets	Total
				No. of accessions			
1974	107	–	6,138	3,365	1	–	**9,611**
1975	11,602	–	8,918	4,097	1	–	**24,618**
1976	14,707	–	10,362	4,720	207	–	**29,996**
1977	15,183	–	10,942	5,302	2,633	388	**34,448**
1978	15,599	376	11,199	5,604	6,232	443	**39,453**
1979	17,754	3,505	11,277	6,798	7,432	2,242	**49,008**
1980	19,247	5,905	11,604	7,771	8,432	3,307	**56,266**
1981	20,831	6,791	11,923	7,862	8,744	3,925	**60,076**
1982	21,871	7,447	12,394	8,793	9,959	4,228	**64,692**
1983	22,553	8,904	12,551	8,893	10,436	4,392	**67,729**
1984	24,286	10,321	13,006	9,652	10,886	4,930	**73,081**
1985	26,359	11,093	13,875	9,953	11,456	5,095	**77,831**
1986	27,654	11,765	14,419	10,466	11,820	5,766	**81,890**
1987	29,877	13,243	14,931	10,840	12,182	6,527	**87,600**
1988	30,508	14,003	15,316	11,144	12,858	6,679	**90,508**
1989	31,258	15,024	15,634	11,178	13,396	6,975	**93,465**
1990	32,693	16,499	16,037	11,194	13,536	7,174	**97,133**
1991	33,038	17,820	16,092	11,504	13,853	7,287	**99,594**
1992	33,784	17,993	16,443	11,659	14,235	7,417	**101,531**
1993	34,469	19,195	16,805	11,848	14,383	8,826	**105,526**
1994	35,184	19,887	16,878	12,092	14,644	9,168	**107,853**
1995	35,660	20,903	17,107	12,379	15,002	9,253	**110,304**
1996	36,098	21,191	17,244	12,900	15,307	9,253	**111,993**
1997	36,374	21,264	17,250	13,013	15,342	9,253	**112,496**
1998	36,374	21,290	17,250	13,200	15,342	9,253	**112,709**
1999	36,727	21,392	17,250	13,544	15,342	9,253	**113,508**

[a]As of 1999, 453 of 15,342 groundnut accessions are wild.

tal input costs of the PQU, based on the share of its operations geared to genebank activities. Multifunctional spaces are allocated to the general facility category, computers and peripherals to general capital. Vehicles are leased monthly by ICRISAT so are excluded from the calculations in Table 5.2.

Annual operating costs[1]

Seed storage

Genetic integrity is best achieved by storing the original seeds (or seeds from initial regeneration) in long-term storage. Only two out of three long-term storage rooms are currently in operation, with a total base collection of 44,030 accessions (about 43% of the active holdings). About 1500–2000 seeds

Table 5.2. Capital input costs (US$, 1999 prices) at the ICRISAT genebank.

Cost category	Service life (years)	Replacement cost	Annualized cost[a]
Medium-term storage		**651,531**	**60,764**
Storage facility	40	117,500	5,708
Storage equipment	10	418,645	49,630
Other equipment	10	3,543	420
Seed container	50	111,843	5,006
Long-term storage		**256,969**	**25,258**
Storage facility	40	54,000	2,623
Storage equipment	10	176,877	20,969
Other equipment	10	6,757	801
Seed container	50	19,335	865
Viability testing		**65,340**	**6,067**
Viability-testing facility	40	24,000	1,166
Seed germinator	10	19,780	2,345
Other equipment	10	21,560	2,556
Regeneration		**184,108**	**17,243**
Greenhouse	10	39,134	4,639
Seed-cleaning equipment	10	9,874	1,171
Seed-drying facility	40	18,000	874
Seed-drying equipment	10	60,040	7,118
Seed-processing facility	40	47,500	2,308
Seed-processing equipment	10	9,560	1,133
Seed-health testing		**59,707**	**6,071**
Seed-health testing facility	40	18,018	875
Greenhouse	10	2,789	331
Lab/office equipment	10	36,300	4,303
Computers	5	2,600	562
General capital		**148,840**	**15,634**
General facility	40	87,000	4,226
Office equipment	10	20,000	2,371
Computers	5	41,840	9,037
Total capital cost		**1,366,495**	**131,037**

Note: See Appendix B for further details.
[a]Calculated at a 4% interest rate using equation (3) in Appendix A.

are stored for genetically homogeneous accessions, such as chickpea and groundnut, and about 4000–12,000 seeds are stored for genetically heterogeneous accessions, such as sorghum, pearl millet and pigeonpea. The active collection is maintained for distribution purposes. The maximum size of seed samples per accession in active collections is about 1.5 kg for groundnut and 400 g for other crops.

The genebank manager and a scientific officer spend about 20% of their time managing the storage operation (15% for the active collection and 5% for the base collection). A genebank technician regularly monitors the temperature and relative humidity of the storage rooms, and an engineer at FESP periodically checks the overall conditions of the facility. We allocated one technician's labour for the annual operation and monitoring of the storage facility. Electricity is expensive in India and the multiple module system for storage makes electricity the most costly variable input apart from labour.[2] On average, running one storage module costs about $10,000 per year; the total electricity cost for the storage facility is $80,000 (Table 5.3).

During the period of our study, ICRISAT's genebank was undergoing substantial restructuring of its space allocation to manage and operate the storage facility more efficiently. The plan is to eventually store all accessions in the long-term facility and maintain a smaller working collection in medium-term storage. Plans include closing two large, medium-term storage rooms that expend high levels of both management labour (because they are separately located) and electricity (because of an outdated system). Once the conversion is complete, three medium-term and four long-term storage rooms will remain in the genebank building (down from six and three, respectively). Transferring collections to the base collection has been a major genebank operation over the past few years.

Viability testing

In addition to standard viability testing when seeds are acquired and first regenerated, seeds are tested during storage at regular intervals of 3–10 years for the active collection and 5–20 years for the base collection, depending on the crop type and initial viability rate. Some crops, such as groundnut, lose viability rapidly and require more frequent testing, while some wild species have consistently low viability and also require more regular testing.

ICRISAT's genebank is in the process of testing the viability of all the holdings. A total of 76,000 accessions in medium-term storage were tested during 1997–1999, and the remaining samples were tested during 2000. In 1999, about 25,000 accessions from the active collection were tested for baseline information (mostly sorghum) and 5000 newly harvested materials were tested prior to storage. Two full-time research technicians and a half-time, non-technical staff member were assigned to accommodate this unusual volume of testing.

Regeneration and characterization

Regeneration is the most critical operation in germplasm conservation, and extra care is taken to preserve the genetic integrity of germplasm accessions during the process. Regeneration activities at ICRISAT's genebank are

Table 5.3. Annual operating costs (US$, 1999 prices) of conservation and distribution at the ICRISAT genebank.

Cost category	Labour	Non-labour	Subtotal	Capital
Acquisition	**633**	**6**	**639**	–
Seed handling	517	5	522	–
Overheads	116	1	117	–
(Number of accessions)			(500)	
Medium-term storage	**6,764**	**84,089**	**90,853**	**60,764**
Storage management	2,826	–	2,826	–
Climate control	2,700	68,700	71,400	–
Overheads	1,238	15,389	16,627	–
(Number of accessions)			(113,550)	
Long-term storage	**2,255**	**24,113**	**26,368**	**25,258**
Storage management	942	–	942	–
Climate control	900	19,700	20,600	–
Overheads	413	4,413	4,826	–
(Number of accessions)			(44,030)	
Viability testing	**13,496**	**2,778**	**16,274**	**6,067**
Viability testing	11,026	2,270	13,296	–
Overheads	2,470	508	2,978	–
(Number of accessions)			(30,000)	
Dissemination	**19,557**	**7,944**	**27,501**	**2,064**
Dissemination management	6,568	–	6,568	–
Seed-health testing	5,984	1,650	7,634	–
Packing/shipping	3,426	4,840	8,266	–
Overheads	3,579	1,454	5,033	–
(Number of accessions)			(21,400)	
Duplication	**906**	**810**	**1,716**	–
Packing/shipping	740	662	1,402	–
Overheads	166	148	314	–
(Number of accessions)			(1,080)	
Information management	**8,916**	**7,283**	**16,199**	–
Database management	7,284	4,950	12,234	–
Other expenses	–	1,000	1,000	–
Overheads	1,632	1,333	2,965	–
General management	**41,861**	**22,154**	**64,015**	**15,634**
Managerial staff	34,200	–	34,200	–
Electricity	–	14,880	14,880	–
Other expenses	–	3,220	3,220	–
Overheads	7,661	4,054	11,715	–
Total operating cost	**94,388**	**149,177**	**243,565**	**109,787**

Note: See Appendix B for further details.

undertaken in the post-rainy season only (October–May) to obtain high-quality seeds with a low incidence of disease. The shorter days during the post-rainy season also induce flowering in photosensitive germplasm accessions, enabling their seed production. Most field operations in the 1999/2000 growing season were geared to increasing seed stock for the base collection.

Three senior germplasm scientists manage the overall regeneration and characterization activities, and a total of four scientific officers implement the operations in the field. The senior scientists spend most of their time on research, not conservation, so we allocated only 20% of their labour to regeneration in Tables 5.4 and 5.5. Scientific officers spend some of their time on dissemination activities, thus we allocated 90% of their labour to conservation activities.

The specific accessions to be regenerated in a given growing season are determined by the genebank manager with help from scientific officers. Accessions with low viability and those for long-term storage are given priority. FESP provides services for the preparation of farmland and the operation of cultivation machinery on a charge-back basis. The farming activities for each crop (seed preparation, planting, plant maintenance, harvesting and so on) are conducted by genebank staff, as discussed further below.

The scientific officer in charge of each crop undertakes the agronomic characterization. Characterization of sorghum, wild groundnut and pigeonpea occurs concurrently with regeneration in the post-rainy season, while pearl millet, groundnut and chickpea are typically characterized during the rainy season, but if necessary are also planted out during the post-rainy season.

Sorghum. During the 1999/2000 season, 5000 accessions were planted for regeneration and characterization on 4 ha during the post-rainy season, and 350 accessions were planted on 0.5 ha for characterization during the rainy season. Among the regenerated accessions, 4230 were for storage in the base collection and the remaining 770 were to increase either seed stocks or seed viability. Because sorghum is a cross-pollinating crop, 'selfing' (or bagging) is an important and highly costly aspect of the operation. Each 'panicle' of a plant is covered with a paper bag for about 3 weeks. A scientific officer manages regeneration and characterization for both sorghum and the small millet varieties, but, given that no activities were undertaken for small millet varieties during the time of our research, we allocated all of this officer's labour cost to sorghum.

Pearl millet. This crop is also regenerated during the post-rainy season, and characterization occurs in both the post-rainy (four traits) and the rainy seasons (20 traits). The number of cultivated accessions varies by year. In 1999/2000, for example, about 1400 accessions were grown in the post-rainy season, predominantly for characterization, and 521 accessions were planted in the rainy season for characterization. For cost calculations, we averaged the numbers of accessions for regeneration and characterization in each season (Table 5.4). Four 4-m rows are planted per accession for regeneration and two for characterization. Thus, the size of the planting area is about 1 ha for both activities (660 accessions for regeneration and 1250 for characterization). Pearl millet is also a vigorously outcrossing plant, so – like sorghum – it requires a highly labour-intensive selfing process when it is regenerated. Cultivation of pearl millet is handled exclusively by a scientific officer assisted by two research technicians and two non-technical staff.

Table 5.4. Annual costs (US$, 1999 prices) of regeneration and characterization at the ICRISAT genebank: sorghum, pearl millet and pigeonpea.

Cost category	Sorghum				Pearl millet				Pigeonpea			
	Labour	Non-labour	Subtotal	Capital	Labour	Non-labour	Subtotal	Capital	Labour	Non-labour	Subtotal	Capital
Regeneration	**64,543**	**18,746**	**83,289**	**8,255**	**19,174**	**7,608**	**26,782**	**1,090**	**17,279**	**12,017**	**29,296**	**991**
Field operation	**29,425**	**6,174**	**35,599**	—	**14,437**	**5,948**	**20,385**	—	**13,066**	**10,508**	**23,574**	—
Field management	6,110	—	6,110	—	5,480	—	5,480	—	4,220	—	4,220	—
Seed preparation	640	50	690	—	280	7	287	—	280	6	286	—
Land preparation	2,180	660	2,840	—	706	183	889	—	565	122	687	—
Planting	640	—	640	—	280	—	280	—	496	—	496	—
Plant maintenance	2,800	446	3,246	—	1,508	7	1,515	—	474	54	528	—
Selfing[a]	7,652	1,996	9,648	—	1,884	4,267	6,151	—	1,760	7,560	9,320	—
Harvesting/threshing	3,208	1,270	4,478	—	1,360	168	1,528	—	1,980	152	2,132	—
Miscellaneous costs	810	622	1,432	—	297	228	525	—	900	691	1,591	—
Overheads	5,385	1,130	6,515	—	2,642	1,088	3,730	—	2,391	1,923	4,314	—
Seed processing	**35,118**	**12,572**	**47,690**	**8,255**	**4,737**	**1,660**	**6,397**	**1,090**	**4,213**	**1,509**	**5,722**	**991**
Process management	2,809	—	2,809	1,710	371	—	371	226	337	—	337	205
Seed cleaning	1,195	—	1,195	582	158	—	158	77	143	—	143	70
Medium-term packing	700	—	700	—	175	—	175	—	84	—	84	—
Seed drying	537	4,771	5,308	3,972	71	630	701	524	64	573	637	477
Long-term seed-health testing	19,950	5,500	25,450	1,991	2,633	726	3,359	263	2,394	660	3,054	239
Long-term packing[b]	3,500	—	3,500	—	462	—	462	—	420	—	420	—
Overheads	6,427	2,301	8,728	—	867	304	1,171	—	771	276	1,047	—
(Number of accessions)	*(5,000)*		*(5,000)*		*(660)*		*(660)*		*(600)*		*(600)*	
Characterization												
Post-rainy season	**7,913**	**255**	**8,168**	—	**13,486**	**1,068**	**14,554**	—	**2,913**	—	**2,913**	—
Field operation	4,847	—	4,847	—	5,629	534	6,163	—	2,913	—	2,913	—
Recording traits	3,960	—	3,960	—	3,409	436	3,845	—	2,380	—	2,380	—
Overheads	887	—	887	—	1,030	98	1,128	—	533	—	533	—
(Number of accessions)	*(5,000)*		*(5,000)*		*(1,250)*		*(1,250)*		*(600)*		*(600)*	
Rainy season	**3,066**	**255**	**3,321**	—	**7,857**	**534**	**8,391**	—	—	—	—	—
Field operation	2,010	208	2,218	—	3,409	436	3,845	—	—	—	—	—
Recording traits	495	—	495	—	3,010	—	3,010	—	—	—	—	—
Overheads	561	47	608	—	1,438	98	1,536	—	—	—	—	—
(Number of accessions)	*(350)*		*(350)*		*(1,250)*		*(1,250)*					

Note: See Appendix B for further details.

[a] 'Selfing' is the process by which cross-pollination of sorghum is minimized through placing paper bags over each 'panicle' of each plant.

[b] All regenerated accessions are assumed to have been tested and packed for long-term storage.

Table 5.5. Annual costs (US$, 1999 prices) of regeneration and characterization at the ICRISAT genebank: chickpea, groundnut and wild groundnut.

Cost category	Chickpea				Groundnut				Wild groundnut			
	Labour	Non-labour	Subtotal	Capital	Labour	Non-labour	Subtotal	Capital	Labour	Non-labour	Subtotal	Capital
Regeneration	**33,371**	**7,923**	**41,294**	**3,963**	**24,108**	**4,471**	**28,579**	**2,147**	**7,402**	**7,139**	**14,541**	**4,804**
Field operation	**16,515**	**1,889**	**18,404**	–	**13,610**	**1,202**	**14,812**	–	**6,583**	**6,888**	**13,471**	**4,639**
Field management	4,850	–	4,850	–	3,924	–	3,924	–	1,556	–	1,556	–
Seed preparation	520	24	544	–	760	13	773	–	280	–	280	–
Land preparation	1,192	271	1,463	–	833	237	1,070	–	742	286	1,028	–
Planting	1,080	–	1,080	–	800	–	800	–	280	–	280	–
Plant maintenance	2,296	120	2,416	–	1,842	126	1,968	–	1,400	5,216	6,616	–
Harvesting/threshing	2,880	610	3,490	–	2,600	330	2,930	–	1,120	25	1,145	–
Miscellaneous costs	675	518	1,193	–	360	276	636	–	–	100	100	–
Overheads	3,022	346	3,368	–	2,491	220	2,711	–	1,205	1,261	2,466	–
Seed processing	**16,856**	**6,034**	**22,890**	**3,963**	**10,498**	**3,269**	**13,767**	**2,147**	**819**	**251**	**1,070**	**165**
Process management	1,348	–	1,348	821	730	–	730	445	56	–	56	34
Seed cleaning	573	–	573	279	336	–	336	151	35	–	35	12
Medium-term packing	336	–	336	–	364	–	364	–	28	–	28	–
Seed drying	258	2,290	2,548	1,907	140	1,241	1,381	1,033	11	95	106	79
Long-term seed-health testing	9,576	2,640	12,216	956	5,187	1,430	6,617	518	399	110	509	40
Long-term packing	1,680	–	1,680	–	1,820	–	1,820	–	140	–	140	–
Overheads	3,085	1,104	4,189	–	1,921	598	2,519	–	150	46	196	–
(Number of accessions)			*(2,400)*				*(1,300)*				*(100)*	
Characterization	**8,600**	**508**	**9,108**	–	**15,470**	**1,314**	**16,784**	–	**1,756**	–	**1,756**	–
Post-rainy season	**8,600**	**508**	**9,108**	–	**5,385**	**438**	**5,823**	–	**1,756**	–	**1,756**	–
Field operation	4,156	415	4,571	–	2,965	358	3,323	–	–	–	–	–
Recording traits	2,870	–	2,870	–	1,435	–	1,435	–	1,435	–	1,435	–
Overheads	1,574	93	1,667	–	985	80	1,065	–	321	–	321	–
(Number of accessions)			*(2,000)*				*(600)*				*(100)*	
Rainy season	–	–	–	–	**10,085**	**876**	**10,961**	–	–	–	–	–
Field operation	–	–	–	–	5,929	716	6,645	–	–	–	–	–
Recording traits	–	–	–	–	2,310	–	2,310	–	–	–	–	–
Overheads	–	–	–	–	1,846	160	2,006	–	–	–	–	–
(Number of accessions)							*(1,500)*					

Note: See Appendix B for further details.

Pigeonpea. Regeneration and characterization are conducted at the same time in the same field during the post-rainy season. In 1999, a total of 600 pigeonpea accessions were regenerated and characterized on 0.5 ha. Two 9-m rows are planted for each accession, amounting to about 180 plants per accession. Pigeonpea is partially cross-pollinated by insects (such as bees), so once again selfing is necessary. In recent years, muslin cloth bags with about a 2-year lifespan have been used, but 6 m × 9 m polypropylene cages are being considered as an alternative. The net-covered frame cages could be dismantled for reuse for more than 5 years, making them cost-effective despite the high initial set-up cost. Since a scientific officer handles both pigeonpea and wild species, only half his labour costs were allocated to pigeonpea-related activities. The scientific officer is assisted by three research technicians and two non-technical staff. The pigeonpea team manages outcrossing studies, genepool maintenance and wild pigeonpea regeneration, so we allocated only 40% of their labour costs to regeneration and characterization activities.

Chickpea. Both the regeneration and characterization of chickpea are undertaken in the post-rainy season only. In 1999/2000, 2400 accessions were grown for regeneration (2000 for storage in the base collection and 400 for regeneration) and 2000 accessions were grown separately for characterization. For each accession, about 80 plants are grown on a total of 1.5 ha for regeneration, and 40 plants are grown on 0.5 ha for characterization. Chickpea is highly susceptible to wilt disease, so the soil is covered with polyethylene sheets to control soil-borne diseases for at least 6 weeks during the hottest part of the year (May–June). This process is called 'solarization'. One scientific officer manages both chickpea and groundnut, so this labour cost is equally divided between the two.

Groundnut. This crop is characterized in both the post-rainy and rainy seasons, separate from regeneration. In the 1999 post-rainy season, 1300 accessions were planted on 1 ha for regeneration and 600 accessions were planted on 0.5 ha for characterization. In the rainy season, 1500 accessions were planted on 1 ha for characterization. Groundnut plants have a low multiplication rate, so four 4-m rows comprising about 160 plants are planted for each accession. Regeneration requires experienced labour because the correct harvest time is very difficult to estimate, given that the seed parts of the crop remain underground.

Wild groundnut. Two types of wild groundnut are conserved at ICRISAT – seed-producing varieties, conserved in the *ex situ* storage facility, and non-seed-producing varieties, conserved in a field greenhouse. About 453 wild groundnut accessions are conserved in the genebank, 100 or so as a field genebank. Each year, about 100 accessions are regenerated in the greenhouse, and about 60% produce enough seed for conservation. In addition to regeneration, the germplasm scientist maintains the field collection year-round. Each accession is planted in a 24-inch-diameter cement ring. The labour required for maintenance is low, but electricity and water consumption costs are relatively high.

Seed-health testing

A Plant Quarantine Unit (PQU) was set up by the Indian government on the ICRISAT campus to test the health of all outgoing accessions. Of the total of 12,000 accessions shipped from ICRISAT in 1999, 3195 accessions were from the genebank. All incoming samples are examined by the NBPGR, a plant quarantine authority operated by the Indian government to test for exotic diseases and pests. Once cleared of any diseases by the NBPGR, seed samples are then sent to ICRISAT, where they are grown out in a 'post-entry quarantine isolation area' to avoid possible introduction of seed-borne diseases and pests.

In addition to testing the accessions destined for shipment elsewhere, the PQU tests all accessions destined for long-term storage for seed-borne fungi. Instead of the more typical and cheaper blotter test, a more accurate agar-plate method is used to ensure that all germplasm is disease-free. In 1999, about 3000 accessions were tested prior to being added to the base collection.

The PQU is managed by a part-time plant pathologist with three scientific officers, five research technicians and four non-technical staff. Based on the number of accessions treated and the work required for the agar-plate method (see the table notes in Appendix B), we assigned 33% of the PQU's overall expenses to testing the health of the seeds held in long-term storage (Tables 5.4 and 5.5) and 17% to the dissemination activities undertaken by the genebank (Table 5.3) (the balance represents dissemination activities by the breeding programmes).

Seed-acquisition and processing

New seed samples are checked for duplicates to avoid redundancy, recorded in the database and prepared for regeneration if stock is low. The number of new accessions introduced to the ICRISAT genebank over the past few years has been relatively small, and no acquisitions were made in 1999. To derive a cost estimate, we averaged the acquired accessions in recent years and used a typical labour requirement for seed handling.

A substantial portion of the genebank manager's time is allocated to seed processing, involving cleaning, drying, testing viability and packaging. Seed cleaning involves removal of debris, low quality, infested or infected seeds and seeds of different species. Chickpea, pigeonpea and groundnut seeds are cleaned by hand, and sorghum and millet seeds are cleaned by a seed blower. Groundnut also requires shelling the pods before storage, which is highly labour-intensive. After cleaning, seeds are tested for seed viability (as described earlier).

Seeds are first dried in the field or in the seed reception area for several days (or sometimes weeks). This initial drying usually meets medium-term storage requirements, and seeds can be immediately packed in plastic jars for groundnut or aluminium screw-top cans for other crops. The accession number, identity and harvest season are included either on the outside of the packet or inside the can.

For long-term storage, further drying is necessary to ensure viability, so seeds are placed in labelled muslin cloth bags and dried in a seed drying room for several weeks (to reach a moisture content of 3–7%). Electricity to operate the drying module is the major cost for this operation. Since 1998, every accession to be stored in the long-term storage room is tested for seed health by the PQU. Long-term storage samples are packed in variously sized aluminium bags to suit the sample size. Groundnut seeds are double-bagged. Chickpea, pigeonpea and groundnut samples are 200 g; sorghum and pearl millet samples are 75 g; and small millet samples are 25 g. The labour cost associated with packaging seeds for long-term storage is generally much higher than for medium-term storage. Though not all harvested materials are packed both for medium- and long-term storage in practice, we assume that they are packaged for both storage modalities to derive meaningful per-accession seed-processing and regeneration costs in Tables 5.4 and 5.5.

Dissemination and safety duplication

Seed for dissemination and duplication is packed in aluminium bags at a size of 5–6 g for sorghum or millet, about 10–20 seeds for wild species and about 100 seeds for other crops. For some species, such as *Arachis* and *Pennisetum*, live plants are disseminated, involving higher costs. The rhizomes are cut into 15-cm pieces, wrapped in paper towel and polyethylene film, and then put in Jiffy bags for shipment. Accessions disseminated abroad require a Material Transfer Agreement (see Chapter 3), an import permit and in many cases a phytosanitary certificate. The genebank manager's time dedicated to overseeing distribution activities makes up a large part of these costs. (See Table 5.6.)

Preparing samples for safety duplication is similar to the procedures used to prepare them for the base collection, but the minimum sample size is about 25 g for sorghum and millet and 100 g for legumes. Pigeonpea is duplicated at NBPGR and chickpea at ICARDA in Aleppo, Syria. A total of 2525 pigeonpea accessions and 2000 chickpea accessions have been duplicated in each location. At the time of this research, 2691 accessions were waiting for quarantine clearance by the Syrian government. In addition, a total of 5914 accessions of chickpea from ICARDA are duplicated in ICRISAT genebank.

Economic Analysis

Representative annual costs of genebank operation

Table 5.7 consolidates the data from earlier tables to provide an overview of the total capital, quasi-fixed and variable costs of operating the ICRISAT genebank, along with the corresponding average costs per accession. The total cost of conserving and distributing germplasm at ICRISAT in 1999 was estimated at $651,661, which includes conservation and distribution costs but excludes costs related to prebreeding, research or training. Some activities in 1999, such as viability testing and seed processing, were exceptional, so we made adjustments accordingly to derive more representative figures.

Table 5.6. Dissemination of germplasm from the ICRISAT genebank, 1989–1999.

Activity/crop	1989	1990	1991	1992	1993	1994	1995	1996	1997	1998	1999
Number of samples											
Sorghum	**18,908**	**16,048**	**13,318**	**12,295**	**14,980**	**7,924**	**2,983**	**3,525**	**4,667**	**4,731**	**5,456**
Outside ICRISAT	15,699	11,968	10,121	10,647	14,273	7,354	2,886	3,044	3,900	4,022	1,144
Inside ICRISAT	3,209	4,080	3,197	1,648	707	570	97	481	767	709	4,312
Pearl millet	**12,645**	**4,317**	**5,253**	**3,072**	**7,106**	**2,301**	**3,143**	**2,695**	**1,224**	**1,344**	**1,980**
Outside ICRISAT	11,872	3,507	3,291	2,824	5,606	1,617	940	1,568	915	1,330	366
Inside ICRISAT	773	810	1,962	248	1,500	684	2,203	1,127	309	14	1,614
Chickpea	**12,473**	**6,554**	**9,938**	**14,865**	**15,582**	**9,329**	**2,893**	**9,778**	**3,283**	**7,046**	**6,756**
Outside ICRISAT	8,610	2,763	3,487	1,917	2,623	1,144	1,300	5,786	1,414	1,632	2,833
Inside ICRISAT	3,863	3,791	6,451	12,948	12,959	8,185	1,593	3,992	1,869	5,414	3,923
Pigeonpea	**6,722**	**5,273**	**3,690**	**4,266**	**9,630**	**6,520**	**2,206**	**2,866**	**1,014**	**962**	**1,595**
Outside ICRISAT	5,682	3,396	1,252	3,076	4,123	3,634	1,407	1,424	837	856	693
Inside ICRISAT	1,040	1,877	2,438	1,190	5,507	2,886	799	1,442	177	106	902
Groundnut	**5,068**	**4,654**	**1,917**	**4,965**	**7,815**	**6,180**	**3,737**	**3,443**	**4,787**	**2,475**	**5,604**
Outside ICRISAT	3,039	2,739	1,311	2,925	6,469	5,743	3,131	2,564	4,657	2,039	3,011
Inside ICRISAT	2,029	1,915	606	2,040	1,346	437	606	879	130	436	2,593
Total	**55,816**	**36,846**	**34,116**	**39,463**	**55,113**	**32,254**	**14,962**	**22,307**	**14,975**	**16,558**	**21,391**
Number of shipments											
Sorghum	**132**	**158**	**133**	**116**	**113**	**101**	**73**	**56**	**74**	**74**	**54**
Outside ICRISAT	88	98	88	87	77	79	54	38	46	46	31
Inside ICRISAT	44	60	45	29	36	22	19	18	28	28	23
Pearl millet	**75**	**73**	**66**	**62**	**78**	**62**	**55**	**72**	**30**	**30**	**39**
Outside ICRISAT	53	49	43	43	57	40	40	59	23	27	25
Inside ICRISAT	22	24	23	19	21	22	15	13	7	3	14
Chickpea	**114**	**85**	**102**	**95**	**82**	**68**	**81**	**87**	**56**	**79**	**63**
Outside ICRISAT	56	53	56	60	45	42	39	47	28	43	42
Inside ICRISAT	58	32	46	35	37	26	42	40	28	36	21
Pigeonpea	**79**	**90**	**83**	**101**	**86**	**77**	**65**	**76**	**54**	**46**	**54**
Outside ICRISAT	55	65	55	64	51	46	44	51	34	42	36
Inside ICRISAT	24	25	28	37	35	31	21	25	20	4	18
Groundnut	**106**	**88**	**94**	**104**	**90**	**69**	**67**	**79**	**45**	**58**	**51**
Outside ICRISAT	48	31	50	55	46	38	40	47	28	23	29
Inside ICRISAT	58	57	44	49	44	31	27	32	17	35	22
Total	**506**	**494**	**478**	**478**	**449**	**377**	**341**	**370**	**259**	**287**	**261**

Table 5.7. Annual total (US$, 1999 prices) and average (US$ per accession, 1999 prices) costs of each operation at the ICRISAT genebank.

Cost category	Number of accessions	Total capital cost	Total quasi-fixed cost[a]	Total variable cost	Average capital cost	Average quasi-fixed cost	Average variable cost
Medium-term storage	113,550	63,109	9,093	93,792	0.56	0.08	0.83
Long-term storage	44,030	26,040	3,031	27,347	0.59	0.07	0.62
Acquisition	500	313	1,212	1,031	0.63	2.42	2.06
Viability testing	30,000	8,412	9,093	19,214	0.28	0.30	0.64
Dissemination	21,400	5,973	21,252	26,301	0.28	0.99	1.23
Safety duplication	1,080	469	1,819	2,304	0.43	1.68	2.13
Regeneration							
Sorghum	5,000	9,037	25,524	61,774	1.81	5.10	12.35
Pearl millet	660	1,871	10,720	20,072	2.84	16.24	30.41
Pigeonpea	600	1,772	8,983	24,323	2.95	14.97	40.54
Chickpea	2,400	4,744	15,575	29,730	1.98	6.49	12.39
Groundnut	1,300	2,928	10,885	21,705	2.25	8.37	16.70
Wild groundnut	100	5,586	4,106	14,445	55.86	41.06	144.45
Characterization							
Sorghum					0.05	2.94	7.77
Post-rainy season	5,000	122	3,377	2,095	0.02	0.68	0.42
Rainy season	350	9	792	2,573	0.02	2.26	7.35
Pearl millet	1,250	130	4,940	10,282	0.10	3.95	8.23
Pigeonpea	600	130	1,855	1,726	0.22	3.09	2.88
Chickpea	2,000	130	1,855	7,921	0.07	0.93	3.96
Groundnut					0.12	2.61	15.04
Post-rainy season	600	37	861	5,154	0.06	1.43	8.59
Rainy season	1,500	93	1,766	9,673	0.06	1.18	6.45
Wild groundnut	100	130	699	1,726	1.30	6.99	17.26
Total cost		**131,035**	**137,438**	**383,188**			

Note: Information and management costs are allocated according to the following percentages for the ICRISAT genebank: medium-term storage (15%), long-term storage (5%), acquisition (2%), viability testing (15%), regeneration (30%), characterization (5%), duplication (3%) and dissemination (25%).
[a]Total quasi-fixed cost includes the cost of senior scientific and technical staff.

Figure 5.1 shows a breakdown of the adjusted annual operating costs. About 45% of the annual operating expenses of the genebank consist of labour costs, which is smaller than the corresponding labour-cost shares we found at other CG genebanks (about two-thirds at ICARDA and CIMMYT, for example). One explanation is the significantly lower average labour cost in India, especially for highly skilled scientists, compared with other regions. Figure 5.1 also indicates a relatively high share of non-labour operating costs (34%, compared with 13% for CIMMYT or ICARDA) as a result of high energy costs in India. The share of capital cost is small (21%), and similar to that in other genebanks.

Economic costs

Annual average costs

The three right-hand side columns in Table 5.7 indicate the average costs of various operations for various crops. The average cost of holding over an accession of any crop for 1 more year, without regeneration, including both variable and quasi-fixed costs, is just 69 cents (Table 5.8). If regeneration is needed, the cost jumps to between $9.89 for sorghum and $93.92 for wild groundnut (which, as previously mentioned, requires extra care). The cost of storing a newly introduced accession in its first year is between $19.14 and $103.17 per accession.

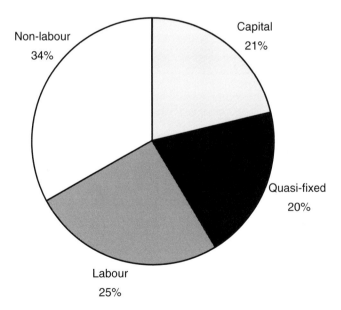

Fig. 5.1. Representative annual costs of maintaining the ICRISAT genebank holdings. Share of conservation and distribution costs by cost category (US$651,661, 1999 prices).

Table 5.8. Average costs (US$ per accession, 1999 prices) of conserving and distributing an accession for 1 year at the ICRISAT genebank.

Cost category	Existing accession		New accession (with regeneration)
	Without regeneration	With regeneration	
Conservation			
Long-term storage	0.69	0.69	0.69
New introduction			
Acquisition			4.49
Initial viability testing			0.94
Initial duplication			3.82
Viability testing		0.47	0.47
Regeneration[a]			
Sorghum		8.73	8.73
Pearl millet		23.33	23.33
Pigeonpea		27.76	27.76
Chickpea		9.44	9.44
Groundnut		12.53	12.53
Wild groundnut		92.76	92.76
Conservation cost			
Sorghum	**0.69**	**9.89**	**19.14**
Pearl millet	**0.69**	**24.49**	**33.74**
Pigeonpea	**0.69**	**28.92**	**38.17**
Chickpea	**0.69**	**10.60**	**19.85**
Groundnut	**0.69**	**13.69**	**22.94**
Wild groundnut	**0.69**	**93.92**	**103.17**
Distribution			
Medium-term storage	0.91	0.91	0.91
Dissemination	2.22	2.22	2.22
Viability testing		0.47	0.47
Regeneration			
Sorghum		8.73	8.73
Pearl millet		23.33	23.33
Pigeonpea		27.76	27.76
Chickpea		9.44	9.44
Groundnut		12.53	12.53
Wild groundnut		92.76	92.76
Characterization			
Sorghum			10.71
Pearl millet			12.18
Pigeonpea			5.97
Chickpea			4.89
Groundnut			17.65
Wild groundnut			24.25
Distribution cost			
Sorghum	**3.13**	**12.33**	**23.04**
Pearl millet	**3.13**	**26.93**	**39.11**
Pigeonpea	**3.13**	**31.36**	**37.33**
Chickpea	**3.13**	**13.04**	**17.93**
Groundnut	**3.13**	**16.13**	**33.78**
Wild groundnut	**3.13**	**96.36**	**120.61**

[a]Regeneration costs are equally allocated between conservation and distribution.

The costs of distributing accessions include medium-term storage and packaging and shipping. The cost of distributing an accession is $3.13 if regeneration is not necessary, and ranges from $12.33 for sorghum to $96.36 for wild groundnut if regeneration is necessary. Distributing a newly acquired accession (which includes characterization) ranges from $17.93 to $120.61 for the crops in the ICRISAT genebank. Notably, the cost of distributing accessions is greater than the cost of conserving them long-term because distribution costs include the comparatively higher costs of storing seeds over the medium term as well as the costs of packing and shipping seeds.

Average costs in the long run

Table 5.9 provides the present values of the costs of conserving and distributing an accession in perpetuity (in constant, inflation-adjusted terms). Once again, we used a 4% baseline interest rate and assumed that viability testing begins in the 10th year after acquisition, that retesting occurs every 5 years thereafter and that an accession is disseminated once every 5 years. On this basis, the average cost of conserving an existing accession in perpetuity ranges from $22.26 for sorghum to $36.02 for wild groundnut. A newly acquired accession costs from $40.71 to $138.49 per accession. The present value of the distribution cost on existing accessions ranges from $45.58 and $132.70 if stocks are sufficient to negate the need for regeneration. If the accession is newly acquired, the present-value distribution cost increases to between $65.49 and $250.18 per accession.

Total costs in the long run

To underwrite the capital costs of conserving and distributing ICRISAT's current collection in perpetuity, the present-value equivalent of $2,084,753 and $1,992,390, respectively, is needed. Excluding the cost of capital, the labour and operating costs to conserve and distribute the entire holdings in perpetuity amount to the present-value equivalents of $5,144,032 and $8,384,360, respectively (Table 5.10). Aggregating these costs, we estimate that $7,228,785 and $10,376,750 are needed, respectively, to conserve and distribute the current level of holdings in the ICRISAT genebank in perpetuity. The total amount needed to be set aside at 4% real interest to generate an annual revenue stream sufficient to meet these annual costs is $17,605,535 – again, a sizeable but not especially large sum of money.

Notes

[1]Kameswara Rao and Bramel (2000) discuss the details of the genebank operations.
[2]In fact, electricity accounts for more than 70% of the annual non-labour operating costs in ICRISAT, compared with just about 15% for CIMMYT or ICARDA.

Table 5.9. Present values (US$ per accession, 1999 prices) of conserving and distributing an accession in perpetuity at the ICRISAT genebank.

Cost category	Existing accession[a]			New accession		
	2%	4%	6%	2%	4%	6%
Conservation						
Long-term storage	35.19	17.94	12.19	35.19	17.94	12.19
New introduction						
Acquisition				4.49	4.49	4.49
Initial viability testing				0.94	0.94	0.94
Initial duplication				3.82	3.82	3.82
Viability testing[b]	4.53	2.18	1.39	4.53	2.18	1.39
Safety duplication[c]	2.26	0.63	0.22	2.26	0.63	0.22
Regeneration (50 years)						
Sorghum	5.44	1.51	0.53	14.64	10.71	9.73
Pearl millet	14.07	3.90	1.37	37.87	27.70	25.16
Pigeonpea	16.69	4.62	1.62	44.91	32.85	29.85
Chickpea	5.86	1.62	0.57	15.77	11.53	10.48
Groundnut	7.69	2.13	0.75	20.70	15.14	13.75
Wild groundnut	55.11	15.27	5.35	148.34	108.49	98.58
Conservation cost						
Sorghum	**47.42**	**22.26**	**14.33**	**65.87**	**40.71**	**32.78**
Pearl millet	**56.05**	**24.65**	**15.17**	**89.10**	**57.70**	**48.21**
Pigeonpea	**58.67**	**25.37**	**15.42**	**96.14**	**62.85**	**52.90**
Chickpea	**47.84**	**22.37**	**14.37**	**67.00**	**41.53**	**33.53**
Groundnut	**49.67**	**22.88**	**14.55**	**71.93**	**45.14**	**36.80**
Wild groundnut	**97.09**	**36.02**	**19.15**	**199.57**	**138.49**	**121.63**
Distribution						
Medium-term storage	46.21	23.56	16.01	46.21	23.56	16.01
Dissemination[d]	23.57	12.48	8.79	23.57	12.48	8.79
Regeneration (25 years)						
Sorghum	23.29	9.54	5.06	32.49	18.74	14.26
Pearl millet	60.23	24.68	13.09	84.03	48.47	36.89
Pigeonpea	71.44	29.27	15.53	99.67	57.49	43.76
Chickpea	25.08	10.28	5.45	34.99	20.19	15.36
Groundnut	32.92	13.49	7.16	45.92	26.49	20.16
Wild groundnut	235.95	96.66	51.29	329.17	189.89	144.52
Characterization						
Sorghum				10.71	10.71	10.71
Pearl millet				12.18	12.18	12.18
Pigeonpea				5.97	5.97	5.97
Chickpea				4.89	4.89	4.89
Groundnut				17.65	17.65	17.65
Wild groundnut				24.25	24.25	24.25
Distribution cost						
Sorghum	**93.07**	**45.58**	**29.86**	**112.98**	**65.49**	**49.77**
Pearl millet	**130.01**	**60.72**	**37.89**	**165.99**	**96.69**	**73.87**
Pigeonpea	**141.22**	**65.31**	**40.33**	**175.42**	**99.50**	**74.53**
Chickpea	**94.86**	**46.32**	**30.25**	**109.66**	**61.12**	**45.05**
Groundnut	**102.70**	**49.53**	**31.95**	**133.35**	**80.18**	**62.61**
Wild groundnut	**305.73**	**132.70**	**76.09**	**423.20**	**250.18**	**193.57**

Note: Some figures in this table were calculated using equations (1) and (2) in Appendix A.
[a]Existing accessions are assumed to have been freshly regenerated. New accessions require initial regeneration.
[b]Viability testing commences in the 10th year after acquisition, and then every 5 years thereafter.
[c]Safety duplication is made concurrently with each round of regeneration.
[d]Dissemination occurs every 5 years.

Table 5.10. Total costs of conservation and distribution in perpetuity at the ICRISAT genebank.

Cost category	Number of accessions	Per-accession cost (US$ per accession, 1999 prices)			Total cost (US$, 1999 prices)		
		Conservation	Distribution	Total	Conservation	Distribution	Total
Sorghum	36,720	**59.83**	**83.40**	**143.23**	**2,196,944**	**3,062,246**	**5,259,190**
Non-capital		40.70	65.49	106.19	1,494,345	2,404,693	3,899,038
Capital		19.13	17.91	37.04	702,599	657,553	1,360,152
Pearl millet	21,400	**77.41**	**115.70**	**193.11**	**1,656,677**	**2,475,927**	**4,132,604**
Non-capital		57.68	96.69	154.37	1,234,409	2,069,118	3,303,527
Capital		19.73	19.01	38.74	422,268	406,809	829,077
Pigeonpea	13,550	**82.64**	**118.74**	**201.38**	**1,119,740**	**1,608,972**	**2,728,712**
Non-capital		62.84	99.50	162.34	851,435	1,348,225	2,199,660
Capital		19.80	19.24	39.04	268,305	260,747	529,052
Chickpea	17,250	**60.75**	**79.21**	**139.96**	**1,047,987**	**1,366,317**	**2,414,304**
Non-capital		41.52	61.11	102.63	716,226	1,054,159	1,770,385
Capital		19.23	18.10	37.33	331,761	312,158	643,919
Groundnut	14,800	**64.51**	**98.62**	**163.13**	**1,112,904**	**1,701,094**	**2,813,998**
Non-capital		45.12	80.18	125.30	778,377	1,383,077	2,161,454
Capital		19.39	18.44	37.83	334,527	318,017	652,544
Wild groundnut	500s	**189.07**	**324.39**	**513.46**	**94,533**	**162,194**	**256,727**
Non-capital		138.48	250.18	388.66	69,240	125,088	194,328
Capital		50.59	74.21	124.80	25,293	37,106	62,399
All crops	**104,220**	**534.21**	**820.06**	**1354.27**	**7,228,785**	**10,376,750**	**17,605,535**

6 IRRI Genebank

BONWOO KOO, PHILIP G. PARDEY AND MICHAEL T. JACKSON

History of the International Rice Genebank at IRRI

The rice genetic resources stored in the International Rice Genebank (IRG) at the International Rice Research Institute (IRRI) at Los Baños, the Philippines, represents the largest and most diverse *ex situ* collection of rice in the world (Jackson, 1997). The collection, estimated at more than 100,000 samples in 1999, is made up of landraces nurtured by farmers for generations, modern and obsolete varieties, some breeding lines and special genetic stocks, the 22 wild species in the genus *Oryza* and related genera in the tribe *Oryzeae*.

National and regional attempts were made during the 1950s to conserve rice genetic resources, including attempts to set up regional collections at three locations in Asia by FAO's International Rice Commission. After the establishment of a rice genebank within IRRI in the early 1960s, many countries donated duplicate samples for long-term storage there. Some of these efforts led to the collection of important genetic materials, such as the Assam Rice Collection, with more than 6600 samples assembled from north-east India in the late 1960s. Another set of germplasm was the deep-water varieties collected in Cambodia before the Khmer Rouge banned their cultivation. From the 1970s, the IRRI genebank began coordinating the collection of rice germplasm worldwide and the size of the collection has increased significantly since then. Between 1961 and 1974, the collection accumulated 28,000 accessions; by 1980 the collection had doubled, and then it almost doubled again in the next 20 years.

Between 1995 and 2000, with support from the Swiss Agency for Development and Cooperation (SDC), IRRI coordinated a major collection programme in 23 countries in South and South-east Asia, sub-Saharan Africa and Costa Rica. The aim of these joint collection activities was to complete, as far as possible, the field collection of cultivated rice germplasm by 2000, to explore areas that had previously been inaccessible (primarily for reasons of civil unrest, which made collecting unsafe) and to enlarge the pool of wild germplasm in *ex situ* conservation. A particular effort was made to collect germplasm in the Lao People's Democratic Republic, a country where much of the existing rice cultivation (often of glutinous varieties) is based on indigenous rice, with thousands of different varieties (Appa Rao *et al.*, 1997). The Lao germplasm is now the second-largest component of the IRRI genebank collection.

The present genebank facilities were constructed in 1977 to establish the International Rice Germplasm Center (renamed as the International Rice Genebank in 1995) and became part of the Genetic Resources Center in 1990. Following extensive renovation and expansion in the early 1990s, significant improvements were made to ensure the long-term security of the germplasm collections. Today, IRG has two storage rooms (long- and medium-term), with a combined capacity for about 120,000 accessions (Table 6.1).

Costing the IRRI Genebank

Capital input costs

A breakdown of the capital costs related to the genebank facility and the costs of the operating equipment used by the IRRI genebank is provided in Table 6.2. The long-term (57-m^2) storage room is located in the middle of the medium-term storage room (whose net size is 182 m^2). Seeds for long-term

Table 6.1. IRRI genebank holdings, 1973–2001.

Year	Total number of accessions	Year	Total number of accessions
1973	24,162	1988	76,297
1974	26,818	1989	77,061
1975	30,332	1990	77,075
1976	34,229	1991	78,381
1977	36,956	1992	n/a
1978	40,768	1993	n/a
1979	47,743	1994	n/a
1980	53,431	1996	n/a
1981	57,027	1995	80,797
1982	60,181	1997	81,585
1983	63,490	1998	84,247
1984	64,744	1999	86,805
1985	66,836	2000	n/a
1986	71,878	2001[a]	91,025
1987	74,498		

[a]Includes only registered accessions as of August 2001.
n/a, not available.

Table 6.2. Capital input costs (US$, 1999 prices) at the IRRI genebank.

Cost category	Service life (years)	Replacement cost	Annualized cost[a]
Medium-term storage		345,232	28,965
Storage facility	40	150,440	7,308
Storage equipment	10	167,180	19,819
Heat-sealing device	10	2,440	289
Seed container	25	25,172	1,549
Long-term storage		195,300	16,728
Storage facility	40	56,418	2,741
Storage equipment	10	90,960	10,783
Vacuum-canning device	10	14,350	1,701
Seed container	50	33,572	1,503
Viability testing		147,510	15,774
Viability-testing facility	40	24,480	1,189
Germination chamber	10	106,000	12,566
Other equipment	10	17,030	2,019
Regeneration		616,088	62,236
Farming facility	40	76,500	3,716
Farming equipment	10	44,721	5,302
Screenhouse	10	222,042	26,323
Embryo-rescue facility	40	6,120	297
Embryo-rescue equipment	10	14,799	1,754
Seed-processing facility	40	65,280	3,171
Seed-cleaning equipment	10	16,350	1,938
Seed-drying facility	40	20,740	1,008
Seed-drying equipment	10	125,536	14,882
Vehicles	7	24,000	3,845
Seed-health testing		37,601	3,422
Seed-health testing facility	40	17,326	842
Lab/office equipment	10	17,503	2,075
Computer	5	1,092	236
Vehicle	7	1,680	269
General capital		224,960	27,692
General facility	40	103,360	5,021
Office equipment	10	30,000	3,556
Computer	5	79,600	17,193
Vehicle	7	12,000	1,922
Total capital cost		**1,566,691**	**154,817**

Note: See Appendix B for details.
[a]Calculated at a 4% interest rate using equation (3) in Appendix A.

storage are kept in vacuum-sealed aluminium cans, and seeds for medium-term storage are stored in heat-sealed aluminium-foil packets. Viability testing is performed year-round with a total of five germination chambers. A screenhouse of 4000 m^2 is used to cultivate wild varieties or those of low viability or in low stock. In addition, some dormant seeds and samples with particularly small quantities of seed are first seeded in an embryo-rescue room before being planted outside. All incoming and outgoing accessions are tested for seed health by the Seed Health Unit (SHU), which occupies an area of 364 m^2, including reception, incubation and inspection areas. Annualized capital costs are shown in the right-hand column of Table 6.2, calculated using a 4% baseline interest rate.

Annual operating costs[1]

Seed storage
As of 1999, the total number of registered accessions of rice in the collection was 86,805, with a further 20,000 samples being multiplied prior to incorporating into the collection. About 94% of this collection consists of *Oryza sativa*, the rest are *Oryza glaberrima* (1.5%) and wild species (4.5%). Most accessions of seeds are stored both medium term as the active collection and long term as the base collection. For medium-term storage, materials are packed in aluminium-foil bags and stored at 2°C with a relative humidity of 30–35%. Materials for temporary storage (for example, planting or freshly harvested materials) are also stored in the medium-term storage room, in paper bags. The base collection is stored in vacuum-sealed cans maintained at about −18 to −20°C.

As for the other genebank operations at other centres described in earlier chapters, the variable cost of operating the storage facilities consists primarily of labour and electricity. Under the supervision of the genebank manager, the storage area's environment is monitored daily by a genebank technician. With an automatic power-supply system, the average time of running the storage equipment is 8 h/day, and the annual cost of electricity for storage amounts to $15,668 (Table 6.3).[2]

Viability testing
Existing stored seed, freshly regenerated seed and newly acquired seed are all tested for viability. Newly regenerated samples must retain a 90% or higher viability rate to be placed into long-term storage; if the germination of existing accessions falls below 85% of the initial rate, the accession is regenerated in the next growing season. A regular seed-testing schedule was established for all the species in IRRI's genebank based on the existing data set (Naredo *et al.*, 1998).

To test each accession, two 100-seed samples are placed in an oven at 50°C to break dormancy and then put on moist paper towels in a germination chamber with temperatures alternating between 30°C and 20°C under 99% relative humidity for a week before observation.[3] In a typical year, about 15,000 accessions from the storage rooms are tested, and an additional 7000–8000 freshly multiplied accessions are also tested before storage.[4] The amount of incoming material varies by year, but in the sample year of 1999 a total of 4950 accessions were newly introduced to IRG. The total number of genebank accessions tested for viability in 1999 was estimated to be 29,250. The viability testing is conducted by one technician with two contracted workers engaged year-round.

Table 6.3. Annual operating costs (US$, 1999 prices) of conservation and distribution at the IRRI genebank.

Cost category	Labour	Non-labour	Subtotal	Capital
Acquisition	**11,719**	**894**	**12,613**	**1,198**
Seed-health testing	4,435	588	5,023	–
Seed handling	5,226	149	5,375	–
Overheads	2,058	157	2,215	–
(Number of accessions)			*(4,950)*	
Medium-term storage	**6,442**	**10,860**	**17,302**	**28,966**
Storage management	2,311	–	2,311	–
Climate control	3,000	8,953	11,953	–
Overheads	1,131	1,907	3,038	–
(Number of accessions)			*(86,080)*	
Long-term storage	**4,295**	**8,145**	**12,440**	**16,728**
Storage management	1,541	–	1,541	–
Climate control	2,000	6,715	8,715	–
Overheads	754	1,430	2,184	–
(Number of accessions)			*(83,930)*	
Viability testing	**10,142**	**437**	**10,579**	**15,774**
Viability testing	8,361	360	8,721	–
Overheads	1,781	77	1,858	–
(Number of accessions)			*(29,250)*	
Dissemination	**16,135**	**5,422**	**21,557**	**2,224**
Dissemination management	2,696	–	2,696	–
Seed-health testing	8,236	1,092	9,328	–
Packing/shipping	2,370	3,378	5,748	–
Overheads	2,833	952	3,785	–
(Number of accessions)			*(6,200)*	
Duplication	**4,665**	**2,552**	**7,217**	**–**
Packing/shipping	3,846	2,104	5,950	–
Overheads	819	448	1,267	–
(Number of accessions)			*(9,450)*	
Information management	**26,589**	**5,053**	**31,642**	**–**
Database management	21,920	1,766	23,686	–
Other expenses	–	2,400	2,400	–
Overheads	4,669	887	5,556	–
General management	**91,691**	**23,180**	**114,871**	**27,693**
Managerial staff	75,590	–	75,590	–
Electricity	–	5,110	5,110	–
Other expenses	–	14,000	14,000	–
Overheads	16,101	4,070	20,171	–
Total operating cost	**171,678**	**56,543**	**228,221**	**92,583**

Note: See Appendix B for further details.

Regeneration and characterization

To obtain high-quality seeds, regeneration is undertaken only in the dry season, between November and May, when conditions are most conducive for producing high-quality seeds with high viability. This season offers high solar radiation, lower night temperatures and lower pest and disease pressure, as well as short days, which promote flowering in photoperiod-sensitive varieties (Ellis and Jackson, 1995; Kameswara Rao and Jackson, 1996).

Cultivated rice is regenerated in a 10-ha field in the IRRI experiment station. In 1999, a total of 7300 accessions were planted for regeneration, of which 4600 were for the multiplication[5] of incoming materials and 2700 for multiplication or regeneration of existing materials. Most accessions of wild rice require different management practices for seed regeneration. Like other wild species, wild rice requires extra care in its regeneration (such as quarantine). These species, and a few cultivated ones, are regenerated in a screenhouse under the management of a scientist. Each accession is planted in separate seed boxes or pots, and 'selfing' (panicle bagging, as described in Chapter 4) is necessary to minimize outcrossing. Particular care is needed while harvesting wild rice to prevent seed loss from shattering and seed mixture. All these processes require extra labour and materials, making the overall costs of regenerating an accession of wild rice significantly higher than the cost of regenerating cultivated rice (Table 6.4).

Characterizing cultivated rice species is done during the rainy season when full vegetative characteristics are observed. Most farmers grow rice during the rainy season when the characteristics of species are well revealed. In 1999, 2000 accessions were characterized on a 2-ha area. The procedure of planting for characterization is the same as that for regeneration, except that materials are not harvested and a substantial amount of labour is needed for recording the traits of each accession. Morphological and agronomic characteristics are scored in small field plots during the wet season using a standard descriptor list. Almost 90% of the whole collection has been scored for about 50 morphological and agronomic characters.

Seed processing

After seeds are regenerated in the field and shipped to the seed-processing room, each accession is verified as representing the composition of the original sample and containing sufficient material for storage. Accessions of insufficient size revert to medium-term storage until they can be regrown the following season. Seeds are next dried for about a month in the seed drying room, which has a 9000-kg capacity and is operated at 15°C and 15% relative humidity. Seeds equilibrate slowly to a moisture content of around 6–7% (Jackson, 1997).

Dried seeds are pre-cleaned with blowers and then manually cleaned with sieves in the seed-processing room (at 20°C and 40–50% relative humidity). Seed cleaning is one of the most labour-intensive activities in IRRI's genebank operation; it is estimated that one technician can clean about two to five accessions per day.[5] Cleaned seed samples are divided into prelabelled envelopes for viability testing, seed-health testing, duplicate storage and prepacking, as well as for the active and base collections. Since rice seeds take in moisture during the cleaning process, they go through a final drying for a week.

Table 6.4. Annual costs (US$, 1999 prices) of regeneration and characterization at the IRRI genebank.

Cost category	Cultivated rice				Wild rice			
	Labour	Non-labour	Subtotal	Capital	Labour	Non-labour	Subtotal	Capital
Regeneration	**145,233**	**11,164**	**156,397**	**41,704**	**23,399**	**1,226**	**24,625**	**20,533**
Field operation	**42,260**	**10,722**	**52,982**	**22,050**	**16,063**	**1,196**	**17,259**	**19,187**
Field management	7,056	–	7,056	20,819	6,048	–	6,048	18,366
Seed preparation	400	73	473	–	820	5	825	–
Embryo rescue	1,536	957	2,493	1,231	1,024	638	1,662	821
Land preparation	4,198	1,322	5,520	–	600	–	600	–
Seeding	707	–	707	–	300	–	300	–
Transplanting	2,615	–	2,615	–	700	–	700	–
Pest control	5,511	3,327	8,838	–	450	59	509	–
Irrigation/fertilization	972	2,430	3,402	–	300	184	484	–
Purification	686	–	686	–	800	50	850	–
Harvesting	7,298	730	8,028	–	1,600	50	1,650	–
Threshing/cleaning	3,860	–	3,860	–	600	–	600	–
Overheads	7,421	1,883	9,304	–	2,821	210	3,031	–
Seed processing	**102,974**	**443**	**103,417**	**19,653**	**7,336**	**30**	**7,366**	**1,346**
Process management	7,242	–	7,242	2,968	462	–	462	203
Seed drying/cleaning	75,460	365	75,825	16,685	5,436	25	5,461	1,143
Medium-term packing	730	–	730	–	50	–	50	–
Long-term packing	1,460	–	1,460	–	100	–	100	–
Overheads	18,082	78	18,160	–	1,288	5	1,293	–
(Number of accessions)			*(7,300)*				*(500)*	
Characterization	**10,978**	**1,735**	**12,713**	–	**1,951**	–	**1,951**	–
Field operation	5,034	1,430	6,464	–	–	–	–	–
Recording traits	4,016	–	4,016	–	1,608	–	1,608	–
Overheads	1,928	305	2,233	–	343	–	343	–
(Number of accessions)			*(2,000)*				*(500)*	

Note: See Appendix B for further details.

For the base collection, seeds are then packed in two 60-g-capacity aluminium cans per accession; for the active collection, samples are stored in resealable laminated aluminium-foil bags (about 500 g); and, for safety duplication, 20-g seed samples are packed in small aluminium packets. Two to five 10-g packets are also prepared at this time ready for dissemination. The amount of stored seeds for wild species is much smaller than for cultivated species: about 50 seeds for the base collection and 10–20 seeds for dissemination.

Seed-health testing

All incoming and disseminated samples are examined by the Seed Health Unit (SHU) under the supervision of the Philippine Plant Quarantine Service, Bureau of Plant Industry. Formerly, seeds were tested for health just before dissemination; however, since the mid-1990s, newly regenerated samples are tested before storage, and if these samples are requested they are directly disseminated without extra testing.[6] The motivation for this change in protocol is to store only the highest-quality seeds, thereby saving on storage costs.

Under a part-time manager, four scientists and eight technicians verify the health of seed used by the genebank, the International Network for the Genetic Evaluation of Rice (INGER) activity and the breeding programme. In 1999, the SHU tested 12,195 incoming and 57,720 outgoing accessions, of which 3469 and 6200 accessions were destined for or came from the genebank, respectively. Thus, the genebank's share of the total number of accessions processed by the SHU was estimated at 14% – the amount allocated in our calculations.

Dissemination and safety duplication

Seeds that are disseminated from IRRI come either from the genebank, which distributes mostly landraces and raw materials, or through INGER, which sends élite germplasm to further national breeding objectives (Jackson *et al.*, 2000). Since 1973, more than 775,000 packets of seeds have been disseminated worldwide. Fewer than 17,000 of the accessions registered in the genebank have never been requested (Table 6.5).[7] In addition to a Philippine phytosanitary certificate, accessions disseminated internationally require an import permit from the requesting country. A local phytosanitary certificate is required for shipments within the Philippines.

Ninety-five per cent of IRRI genebank accessions are duplicated at the US National Seed Storage Laboratory in Fort Collins via the black-box method described in Chapter 3. Duplicate storage of African rice is shared between IRRI, the International Institute of Tropical Agriculture (IITA) in Nigeria and the Africa Rice Center (WARDA, formerly the West Africa Rice Development Association) in the Ivory Coast. The seeds destined for backup storage are packed at the same time as those kept in on-site storage and dissemination, and so the labour cost involved in duplicating and disseminating seed is similar.

Table 6.5. Dissemination of germplasm from the IRRI genebank, 1991–1999.

Activity/variety	1991	1992	1993	1994	1995	1996	1997	1998	1999
Number of samples	**31,539**	**16,229**	**25,289**	**25,802**	**15,630**	**10,958**	**5,633**	**6,670**	**6,194**
Oryza sativa									
Dissemination	**31,539**	**14,895**	**22,893**	**24,783**	**14,552**	**8,033**	**4,371**	**5,474**	**5,027**
GRC in IRRI	356	2,522	2,488	2,187	2,343	1,660	816	1,935	2,663
Elsewhere in IRRI	25,596	6,421	11,873	11,915	7,590	4,084	2,200	1,117	674
Outside IRRI	5,587	5,952	8,532	10,681	4,619	2,289	1,355	2,422	1,690
Safety duplication	**4,530**	–	**8,608**	–	**5,435**	–	–	**11,727**	**8,343**
Oryza glaberrima									
Dissemination	–	**31**	**247**	**38**	**220**	**275**	**216**	**216**	**35**
GRC in IRRI	–	1	2	–	200	197	172	191	1
Elsewhere in IRRI	–	3	14	23	3	–	–	–	–
Outside IRRI	–	27	231	15	17	78	44	25	34
Safety duplication	–	–	**88**	–	**24**	–	–	**136**	**262**
Wild rice									
Dissemination	–	**1,303**	**2,149**	**981**	**858**	**2,650**	**1,046**	**980**	**1,132**
GRC in IRRI	–	–	190	–	13	1,867	380	104	439
Elsewhere in IRRI	–	344	130	303	140	124	146	14	32
Outside IRRI	–	959	1,829	678	705	659	520	862	661
Safety duplication	–	–	–	–	**1,802**	–	–	**245**	**842**
Number of shipments	**270**	**292**	**288**	**265**	**232**	**215**	**144**	**176**	**186**
Oryza sativa									
Dissemination	**270**	**241**	**226**	**223**	**186**	**149**	**108**	**128**	**141**
GRC in IRRI	14	16	20	12	14	12	12	12	12
Elsewhere in IRRI	141	105	83	101	82	52	42	48	66
Outside IRRI	115	120	123	110	90	85	54	68	63
Oryza glaberrima									
Dissemination	–	**10**	**15**	**8**	**9**	**11**	**9**	**12**	**7**
GRC in IRRI	–	1	1	–	1	1	3	1	1
Elsewhere in IRRI	–	1	5	3	3	–	–	–	–
Outside IRRI	–	8	9	5	5	10	6	11	6
Wild rice									
Dissemination	–	**41**	**47**	**34**	**37**	**55**	**27**	**36**	**38**
GRC in IRRI	–	–	4	–	2	21	4	2	12
Elsewhere in IRRI	–	10	10	14	12	6	5	2	6
Outside IRRI	–	31	33	20	23	28	18	32	20

GRC, Genetic Resources Centre.

Data and information management

Documentation and data management were significantly enhanced during the 1990s through the development of the International Rice Genebank Collection Information System (IRGCIS), which operates on IRRI's local area network, using Oracle as the software platform. Its value has been further strengthened through integration with the CG's SINGER database. The recent launching of a Web-enabled version of the IRGCIS has expanded access to IRRI's germplasm data.

Economic Analysis

Representative annual costs of genebank operation

The total cost of operating the rice genebank at IRRI is estimated at $578,727 in 1999, including the operational costs related to conservation and distribution. Capital expenses constitute only a quarter of the annual operating costs, and labour accounts for more than 60%, including the costs of scientific and senior technical staff engaged in genebank operations, which we consider as quasi-fixed (labour) costs (Table 6.6 and Fig. 6.1).

Economic costs

Annual average costs

The average cost of holding over an accession of any crop for one more year, if initial regeneration is deemed unnecessary, is just 24 cents; if regeneration is required, the cost jumps to between $13.79 for cultivated rice and $34.09 for wild rice. Keeping a newly introduced accession in its first year costs between $21.70 and $42.00 per accession. Disseminating seed samples, if there is sufficient stock in the active collection, costs $8.66 per accession. If regeneration is needed to boost stocks, distribution costs per accession jump to $22.21 for cultivated rice and $42.51 for wild rice. The cost of disseminating newly acquired sample ranges from $31.50 to $49.34 per accession due to the costs involved in characterizing newly acquired material (Table 6.7).

Average costs in the long run

Table 6.8 provides present values for the costs of conserving and distributing an accession in perpetuity (once again, in constant, inflation-adjusted terms). In keeping with the procedures used for estimating costs at other CG centres, we used a 4% baseline interest rate and assumed that viability testing begins in the 10th year after acquisition, that retesting occurs every 5 years thereafter and – in the case of IRRI – that an accession is disseminated once every 10 years.

Under these assumptions, the average cost of conserving an existing accession (not needing regeneration) in perpetuity at IRRI is $10.58 for culti-

Table 6.6. Annual total (US$, 1999 prices) and average (US$ per accession, 1999 prices) costs of each operation at the IRRI genebank.

Cost category	Number of accessions	Total capital cost	Total quasi-fixed cost[a]	Total variable cost	Average capital cost	Average quasi-fixed cost	Average variable cost
Medium-term storage	86,080	33,120	18,509	20,770	0.38	0.22	0.24
Long-term storage	83,930	18,113	7,104	12,661	0.22	0.08	0.15
Acquisition	4,950	3,967	15,808	11,455	0.80	3.19	2.31
Viability testing	29,250	18,544	12,807	12,423	0.63	0.44	0.42
Dissemination	6,200	7,763	29,786	21,074	1.25	4.80	3.40
Safety duplication	9,450	1,385	6,637	7,906	0.15	0.70	0.84
Regeneration							
Cultivated rice	7,300	48,350	42,880	148,680	6.62	5.87	20.37
Wild rice	500	22,195	16,896	16,520	44.39	33.79	33.04
Characterization							
Cultivated rice	2,000	1,108	9,079	9,494	0.55	4.54	4.75
Wild rice	500	277	2,270	1,146	0.55	4.54	2.29
Total cost		**154,822**	**161,776**	**262,129**			

Note: Information and general management costs are allocated according to the following percentages for the IRRI genebank: medium-term storage (15%), long-term storage (5%), acquisition (5%), viability testing (10%), regeneration (30%), characterization (5%), duplication (5%) and dissemination (20%).
[a]Total quasi-fixed cost includes the cost of senior scientific and technical staff.

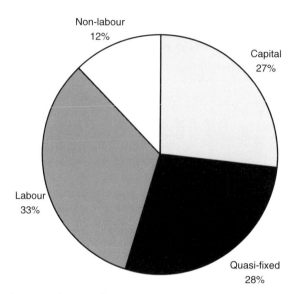

Fig. 6.1. Representative annual costs of maintaining the IRRI genebank holdings. Share of conservation and distribution costs by cost category (US$578,727, 1999 prices).

vated rice and $13.90 for wild rice. For a newly acquired accession or one needing initial regeneration, the conservation costs range from $32.04 to $55.66 per accession. The present value to distribute an accession ranges from $51.19 to $72.23 if an initial regeneration is not required, while distributing a newly acquired accession of rice costs between $74.03 and $112.91.

Total costs in the long run
The present-value equivalent of $1,285,255 and $1,959,936 are needed to underwrite the capital costs of conserving and distributing IRRI's current collection in perpetuity, respectively. Labour and operating costs to conserve and distribute the entire holdings in perpetuity would require a total of $2,857,157 and $6,550,944, respectively, in present values (Table 6.9). We estimate that overall it costs $4,142,412 to conserve and $8,510,880 to distribute the current level of holdings in the IRRI genebank in perpetuity, representing a total of $12,653,292.

Notes

[1]Jackson (2000) discusses the details of the IRRI genebank operations.
[2]This is denoted as 'climate control' costs in Table 6.3. In addition to the use of electricity for storage purposes, the cost of electricity used for general purposes is included as part of the general management category.

Table 6.7. Average costs (US$ per accession, 1999 prices) of conserving and distributing an accession for 1 year at the IRRI genebank.

| Cost category | Existing accession | | New accession[a] (with regeneration) |
	Without regeneration	With regeneration	
Conservation			
Long-term storage	0.24	0.24	0.24
New introduction			
Acquisition			5.51
Initial viability testing			0.86
Initial duplication			1.54
Viability testing		0.43	0.43
Regeneration[b]			
Cultivated rice		13.12	13.12
Wild rice		33.42	33.42
Conservation cost			
Cultivated rice	**0.24**	**13.79**	**21.70**
Wild rice	**0.24**	**34.09**	**42.00**
Distribution			
Medium-term storage	0.46	0.46	0.46
Dissemination	8.20	8.20	8.20
Viability testing		0.43	0.43
Regeneration			
Cultivated rice		13.12	13.12
Wild rice		33.42	33.42
Characterization			
Cultivated rice			9.29
Wild rice			6.83
Distribution cost			
Cultivated rice	**8.66**	**22.21**	**31.50**
Wild rice	**8.66**	**42.51**	**49.34**

[a]Newly introduced accessions require initial regeneration and viability testing.
[b]Regeneration costs are equally allocated between conservation and distribution.

[3]For wild rice, only 20 seeds are tested after breaking dormancy.
[4]The entire collection was tested for viability over several years from the mid-1990s, and more than 600,000 germination tests were made over that time.
[5]The target seed quantity after drying is about 1 kg per accession; assuming 40 seeds weigh 1 g, each accession of cultivated rice consists of 40,000 seeds.
[6]Accessions placed in medium-term storage prior to mid-1990 were not necessarily subject to health testing and thus are tested prior to dissemination.
[7]Some are requested repeatedly, such as IRGC328 (a variety named Azucena), reflecting their common and widespread use in molecular mapping and genomics.

Table 6.8. Present values (US$ per accession, 1999 prices) of conserving and distributing an accession in perpetuity at the IRRI genebank.

Cost category	Existing accession[a]			New accession		
	2%	4%	6%	2%	4%	6%
Conservation						
Long-term storage	12.01	6.12	4.16	12.01	6.12	4.16
New introduction						
Acquisition				5.51	5.51	5.51
Initial viability testing				0.86	0.86	0.86
Initial duplication				1.54	1.54	1.54
Viability testing[b]	4.14	1.99	1.28	4.14	1.99	1.28
Safety duplication[c]	0.91	0.25	0.09	0.91	0.25	0.09
Regeneration (50 years)						
Cultivated rice	8.01	2.22	0.78	21.56	15.77	14.33
Wild rice	20.01	5.54	1.94	53.86	39.39	35.79
Conservation cost						
Cultivated rice	**25.07**	**10.58**	**6.31**	**46.53**	**32.04**	**27.77**
Wild rice	**37.07**	**13.90**	**7.47**	**78.83**	**55.66**	**49.22**
Distribution						
Medium-term storage	23.27	11.86	8.06	23.27	11.86	8.06
Dissemination[d]	45.66	25.28	18.58	45.66	25.28	18.58
Regeneration (25 years)						
Cultivated rice	34.30	14.05	7.46	47.85	27.60	21.01
Wild rice	85.66	35.09	18.62	119.51	68.94	52.47
Characterization						
Cultivated rice				9.29	9.29	9.29
Wild rice				6.83	6.83	6.83
Distribution cost						
Cultivated rice	**103.23**	**51.19**	**34.10**	**126.07**	**74.03**	**56.94**
Wild rice	**154.59**	**72.23**	**45.26**	**195.27**	**112.91**	**85.94**

Note: The data in this table were calculated using equations (1) and (2) in Appendix A.
[a]Existing accessions are assumed to have been freshly regenerated. New accessions require initial regeneration.
[b]Viability testing commences in the 10th year after acquisition, and then every 5 years thereafter.
[c]Safety duplication is made concurrently with each round of regeneration.
[d]Dissemination occurs every 5 years.

Table 6.9. Total costs of conservation and distribution in perpetuity at the IRRI genebank.

Crop/cost category	Number of accessions	Per-accession cost (US$ per accession, 1999 prices)			Total cost (US$, 1999 prices)		
		Conservation	Distribution	Total	Conservation	Distribution	Total
Cultivated rice	83,600	**46.05**	**95.20**	**141.25**	**3,849,174**	**7,958,734**	**11,807,908**
Non-capital		32.05	74.04	106.09	2,679,030	6,189,594	8,868,624
Capital		14.00	21.16	35.16	1,170,144	1,769,140	2,939,284
Wild rice	3,200	**91.63**	**172.54**	**264.17**	**293,238**	**552,146**	**845,384**
Non-capital		55.66	112.92	168.58	178,127	361,350	539,477
Capital		35.97	59.62	95.59	115,111	190,796	305,907
All crops	**86,800**	**137.68**	**267.74**	**405.42**	**4,142,412**	**8,510,880**	**12,653,292**

7 CIAT Genebank

Bonwoo Koo, Philip G. Pardey and Daniel Debouck

History of the CIAT Genebank

Established in Palmira, Colombia, in 1969, Centro Internacional de Agricultura Tropical (the International Center for Tropical Agriculture, known as CIAT) has worked to conserve and increase the productivity of beans, cassava, tropical forages and rice. Four programmes were initially established, with agronomy and breeding as the main components. The animal production programmes (cattle and pigs) were soon phased out, after which CIAT's activities concentrated on tropical pastures only, focusing on seed-based technologies that could be adopted quickly by farmers. Improved germplasm was produced through mass selection from germplasm collections or through plant breeding after the collections were evaluated.

Plant breeding requires large germplasm collections, but not much was known about their diverse agronomic traits or heritability in the 1970s. CIAT quickly accumulated thousands of accessions of beans, cassava and tropical forages as donations from other institutions. As the breeding programme developed, it became necessary to establish and maintain a separate germplasm collection. In 1977, the Genetic Resources Unit (GRU) was formally established (in a building originally designed as a slaughterhouse used by CIAT's animal production programme), and by 1988 the total collection grew from 17,000 to 40,000 accessions. Germplasm explorations for specific materials contributed substantially to the collection over this time.

The inherited facility was gradually modified and improved for seed conservation, and two cold-storage rooms and a drying room were built in 1990 with funding from the Italian government. Countries increasingly recognized the value of well-conserved and evaluated germplasm, and GRU operations continually improved. With the increasing demand for cassava and concern over the safety of disseminated materials, the entire cassava field genebank was progressively converted to an *in vitro* collection during 1980–1985,[1] and

© The International Plant Genetic Resources Institute (for the CGIAR System-wide Genetic Resources Programme) and the International Food Policy Research Institute 2004. *Saving Seeds* (B. Koo, P.G. Pardey and B.D. Wright *et al.*)

protocols for germplasm health testing were initiated (for example, thermo-therapy – a means of heat-treating germplasm to rid it of diseases).

The FAO–CG in-trust agreement of 1994 reinforced the need for GRU to upgrade its operation (as stressed by external CG-genebank reviews in 1995, 1997 and 2000). The upgrades affected three areas: training staff, re-engineering operations and processes and improving facilities. A new, uni-form computer system has also facilitated cross-checking to improve quality control, and a newly installed barcode system has reduced both the time needed to record data and the incidence of errors. Seed-drying and storage capacity has been increased and seed viability and *in vitro* labora-tories have been modernized. The GRU also provided working training for 92 professionals in partner organizations during 1983–2000.

As of 2000, the total number of accessions conserved at the CIAT genebank was estimated at 56,138, including 31,880 accessions of *Phaseolus* beans, 18,178 of tropical forages and 6080 of cassava (clones of *Manihot*) (Table 7.1).[2] For cassava, the entire collection is maintained *in vitro*, as well as in a field genebank as backup (see more detail below). A small number of accessions (130) are also kept under cryoconservation (an experimental process described below). In addition, about 1000 accessions of wild cas-sava are conserved in cold storage as botanical seeds. All *Phaseolus* beans and forages are conserved in cold storage.[3]

Costing the CIAT Genebank

Capital input costs

Cassava

The *ex situ* conservation of cassava germplasm, though it is the only approach guaranteeing access to known traits, is hindered by several basic

Table 7.1. CIAT genebank holdings, 1978–2002.

Year	No. of accessions			
	Cassava	Beans	Forages	Total
1978	50	6,000	300	**6,350**
1980	702	14,626	5,033	**20,361**
1990	4,788	26,506	14,397	**45,691**
1995	5,707	27,918	15,157	**48,782**
1996	5,719	28,393	15,453	**49,565**
1997	5,719	28,623	15,748	**50,090**
1998	5,719	31,656	16,044	**53,419**
1999	5,719	31,880	16,339	**53,938**
2000	5,728	31,880	17,283	**54,891**
2001	5,728	31,881	18,228	**55,837**
2002	5,728	31,881	18,562	**56,171**

characteristics of the plant material – for example, it is highly heterozygous and outcrossing. Currently, four methods are used by the CIAT genebank to conserve cassava germplasm, each with various advantages and disadvantages: *in vitro* conservation of plantlets, field genebank, cold storage of botanical seeds and cryoconservation of shoot tips.

In vitro conservation is amenable to thermotherapy (mentioned above) and disease indexation (a form of health testing described below), so disease-free clones can be distributed internationally in an environment of increasingly strict quarantine requirements. However, *in vitro* collections require subculturing (regeneration in test tubes from plant tips) every 10–18 months to maintain viability. The field genebank is the only method that facilitates germplasm characterization and evaluation, but it is vulnerable to pests and diseases. The conservation of botanical seeds in cold storage can be comparatively cheap, but only two species (*Manihot esculenta* and *Manihot glazovii*) are considered 'orthodox' in that they can be successfully dried for long-term storage.[4] Cryoconservation of shoot tips would allow safe duplication (via the black-box method), but a protocol has yet to be established. Table 7.2 shows the capital costs required for different conservation methods.

Beans and forages

Table 7.3 indicates the capital input items needed for the conservation of bean and forage germplasm in cold storage, categorized by operation. Viability testing is conducted in a separate laboratory next to the GRU. Of note, the greenhouse represents a large capital investment even though most regeneration is done in fields, and Colombia's humidity makes seed-drying equipment all the more important.

Annual operating costs: cassava

Storage and subculturing

For storage, plants of each cassava accession (clones) are placed in five test tubes with a chemical medium and stored in a conservation room set at 23°C and 1500–2000 lux. Under these conditions, plantlets have an effective lifespan of only about 10–18 months (in contrast to the decades-long viability of seeds, as described in earlier chapters). Subculturing (as mentioned above) is a similar process conducted periodically to maintain the viability of each accession. Several shoot tips are extracted from plants and, once again, placed in test tubes with a chemical medium; however, before being stored they are maintained in a growing room for 3 months – a very labour-intensive process. A slow-growth method is currently being developed to increase the interval of subculturing. Table 7.4 shows the costs of storage and subculturing in the CIAT genebank.

Disease indexing

This is a process by which all incoming accessions are tested for various viruses and diseases to comply with the FAO agreement of a full-availability, disease-free collection. The meristem (or tip) of plants is extracted,

Table 7.2. Capital input costs (US$, 2000 prices) for cassava at the CIAT genebank.

Cost category	Service life (years)	Replacement cost			Annualized cost[a]		
		In vitro	Cryoconservation	Field genebank	In vitro	Cryoconservation	Field genebank
Storage		**35,613**	**17,646**	—	**2,708**	**692**	—
Conservation facility	40	21,645	—	—	1,052	—	—
Climate-control equipment	10	3,200	—	—	379	—	—
Cryoconservation equipment	100	—	17,646	—	—	692	—
Other equipment	10	10,768	—	—	1,277	—	—
Subculturing		**83,304**	—	—	**8,361**	—	—
Subculturing facility	40	21,645	—	—	1,052	—	—
Lab equipment	10	52,530	—	—	6,227	—	—
Other equipment	10	9,129	—	—	1,082	—	—
Disease indexation		**116,755**	—	—	**12,526**	—	—
Testing facility	40	21,805	—	—	1,059	—	—
Greenhouse	10	39,000	—	—	4,623	—	—
Lab/office equipment	10	53,783	—	—	6,376	—	—
Computer	5	2,167	—	—	468	—	—
Cryoconservation operation			**19,547**	—		**1,813**	—
Cryoconservation lab facility	40	—	7,215	—	—	351	—
Lab equipment	10	—	12,332	—	—	1,462	—
General capital		**62,188**	**4,969**	**4,969**	**5,747**	**514**	**514**
General facility	40	42,563	3,006	3,006	2,068	146	146
Other equipment	10	5,750	575	575	682	68	68
Computers	5	13,875	1,388	1,388	2,997	300	300
Total capital cost		**297,860**	**42,162**	**4,969**	**29,342**	**3,019**	**514**

Note: See Appendix B for further details.
[a]Calculated at a 4% interest rate using equation (3) in Appendix A.

Table 7.3. Capital input costs (US$, 2000 prices) for conserving bean and forage at the CIAT genebank.

Cost category	Service life (years)	Replacement cost	Annualized cost[a]
Medium-term storage		**232,360**	**17,703**
Storage facility	40	59,484	2,890
Storage equipment	10	52,633	6,240
Other equipment	10	43,243	5,126
Seed containers	50	77,000	3,447
Long-term storage		**143,214**	**12,181**
Storage facility	40	43,130	2,095
Storage equipment	10	44,133	5,232
Other equipment	10	31,843	3,775
Seed containers	50	24,108	1,079
Viability testing		**49,932**	**4,003**
Viability-testing facility	40	27,387	1,330
Testing equipment	10	22,545	2,673
Regeneration		**491,347**	**51,058**
Greenhouse	10	127,200	15,079
Field equipment	10	55,100	6,532
Seed-drying facility	40	18,118	880
Seed-drying equipment	10	108,333	12,843
Seed-processing facility	40	84,656	4,113
Seed-processing equipment	10	97,940	11,611
Seed-health testing		**168,467**	**17,345**
Seed-health testing facility	40	43,567	2,116
Greenhouse	10	13,000	1,541
Lab/office equipment	10	107,567	12,752
Computers	5	4,333	936
General capital		**139,125**	**14,389**
General facility	40	84,175	4,089
Other equipment	10	16,100	1,909
Computers	5	38,850	8,391
Total capital cost		**1,224,445**	**116,679**

Note: See Appendix B for further details.
[a]Calculated at a 4% interest rate using equation (3) in Appendix A.

placed in a medium, heat-treated for 3 weeks (40°C during day and 35°C at night) and grown in a greenhouse for 3 months before being tested. The current protocol involves three tests to detect different types of viruses.[5] Since the majority of the collection was not tested when it was introduced, indexing of all of CIAT's cassava materials began in 1996; the process was completed in 2001. Only two tests (enzyme-linked immunosorbent assay (ELISA) and grafting) were performed as of 2001, and the protocol of the new polymerase chain reaction (PCR) method is still being developed.

Dissemination

Dissemination of cassava is costly because a new set of clones must be sub-cultured each time a request is made. Two to five of the subculture test tubes

Table 7.4. Annual operating costs (US$, 2000 prices) for cassava at the CIAT genebank.

Category	In vitro			Cryoconservation			Field genebank		
	Labour	Non-labour	Capital	Labour	Non-labour	Capital	Labour	Non-labour	Capital
Storage	**3,514**	**7,880**	**2,707**	**836**	**366**	**692**	–	–	–
Management	2,880	–		685	–		–	–	
Electricity/supplies	–	6,459		–	300		–	–	
Overheads	634	1,421		151	66		–	–	
(No. of accessions)			(6,080)			(2,000)			–
Subculturing	**46,248**	**6,739**	**8,361**	**1,748**	**122**	–	–	–	–
Management	11,520	–		–	–		–	–	
Subculturing	26,388	5,524		–	–		–	–	
Viability testing	–	–		1,433	100		–	–	
Overheads	8,340	1,215		315	22		–	–	
(No. of accessions)			(4,290)			(300)			–
Disease indexation	**33,330**	**31,271**	**12,527**	–	–	–	–	–	–
Management	3,890	–		–	–		–	–	
Field operation	7,400	3,532		–	–		–	–	
Lab operation	16,030	22,100		–	–		–	–	
Overheads	6,010	5,639		–	–		–	–	
(No. of accessions)			(1,200)			–			–
Field maintenance	–	–	–	–	–	–	**26,816**	**5,563**	–
Management							7,005	750	
Land preparation							2,570	481	
Planting							2,115	352	
Field maintenance							3,698	2,237	
Harvesting							3,410	740	
Characterization							3,182	–	
Overheads							4,836	1,003	
(No. of accessions)									(5,230)

Cryo-operation	–	–	–	**25,998**	**1,348**	**1,812**	–	–
Management	–	–	–	4,110	–	–	–	–
Processing	–	–	–	17,200	1,105	–	–	–
Overheads	–	–	–	4,688	243	–	–	–
(No. of accessions)	–	–	–	–	–	*(1,000)*	–	–
Dissemination	**16,050**	**2,913**	–	–	–	–	–	–
Management	2,880	–	–	–	–	–	–	–
Subculturing	8,796	1,672	–	–	–	–	–	–
Packing/shipping	1,480	716	–	–	–	–	–	–
Overheads	2,894	525	–	–	–	–	–	–
(No. of accessions)	–	–	*(2,170)*	–	–	–	–	–
General management	**29,152**	**12,091**	**5,746**	**4,274**	**902**	**514**	**902**	**514**
Managerial staff	23,895	–	–	3,503	–	–	–	–
Office expenses	–	9,911	–	–	739	–	739	–
Overheads	5,257	2,180	–	771	163	–	163	–
Total operating cost	**128,294**	**60,894**	**29,341**	**32,856**	**2,738**	**3,018**	**6,465**	**514**

Note: See Appendix B for further details.

are packed and shipped in polystyrene boxes by express courier service. As with the other CG genebanks, a phytosanitary certificate issued by the Colombian authorities and a material transfer agreement are included with the package (Roca *et al.*, 1984). Only indexed accessions are disseminated internationally to comply with quarantine requirements. An average of about 400–500 accessions were disseminated annually in the few years leading up to 2000; 2000 was an exception, with 2176 accessions disseminated in 54 shipments, the majority of which went to Thailand (see Table 7.6 below).

Cryoconservation

Cryoconservation enables long-term storage of cassava germplasm in a reduced space, at low cost, while maintaining its genetic integrity (Escobar *et al.*, 1997). Ongoing research focuses on establishing a basic cryoconservation protocol for cassava – the number and size of shoot tips, types of medium, levels of dehydration and so on. The first step is to propagate plantlets under culture conditions of 26–28°C and 12 hours of light per day (known as the 'photoperiod'). This takes about 2 months, after which the shoot tips are extracted, treated with sodium alginate and made to coagulate into tiny capsules (called beads or 'pseudo-seeds') through immersion in a calcium chloride solution. The beads are put in silica gel for 20 h to reduce the moisture content, and then 30 beads are packed in three small tubes of ten shoot tips each and stored in a tank filled with liquid nitrogen at a temperature of –196°C.[6] In this extreme environment, all biological activity is effectively halted, and the germplasm can be conserved for more than 100 years without maintenance. However, the recovery rate of shoot tips stored at this low temperature is one of the most difficult parts of current research.

As of 2000, 130 accessions were conserved in this way; and the number will increase up to 630 accessions (the size of the core collection at the time of this study). Various freezing methods have been attempted, and the cost data in Table 7.4 reflect the conservation protocols prevailing in 2000. Substantial resources have been allocated to basic research on cryoconservation rather than conservation *per se*, but the data in Tables 7.2 and 7.4 include only conservation-related cost estimates. A small tank (the size of a small refrigerator) can hold 2000 accessions (or 6000 test tubes) of cassava. To generate our storage cost estimates using this technique, we assumed that the tank was full.

Routine storage costs include labour for checking and filling the tank with liquid nitrogen and the material cost of liquid nitrogen.[7] The most expensive element of cryoconservation is preparing the pseudo-seeds before placing them in the storage tank. It is projected that two technicians can prepare about 1000 accessions per year, which is about the same throughput as regenerated seed.

Field genebank

Relying solely on *in vitro* methods – whereby just a few test tubes are used to store each accession – can be unduly risky; thus a field genebank is maintained as a safety backup and to enable characterization and evaluation of accessions. Once a complete evaluation of the collection is made and the protocol for cryoconservation is fully developed, the field genebank may become unnecessary, other than for occasional plantings to screen for new traits or maintain a few élite lines.

As of 2000, a total of 5230 cassava accessions were conserved on 5 ha and managed by the breeding programme in collaboration with the genebank. In 2000, the entire collection was cleared from the field to control chronic pest problems (whiteflies). Fallowing the field for a few months substantially reduces the whitefly population. The replanting cycle has been reduced from 2–3 years down to 1 year to solve the whitefly problem over the long run.[8] The annual operating cost of the field genebank in Table 7.4 is based on the new 1-year planting cycle; however, we did not include the costs of the major land-clearing activity conducted in 2000 because it was an atypical expense. Some of the cassava crops are sold to local companies after characterization or evaluation, but this revenue is also excluded from our cost calculations (about $250/ha).

Annual operating costs: beans and forages

Storage

The medium-term storage room holds 31,880 accessions of beans, 18,178 of forages and 1000 of wild cassava. About 40% of these collections are also stored long-term (12,705 accessions of beans and 7389 accessions of forage). Accessions are stored at the CIAT genebank for the same purposes as at the other genebanks already discussed: as a base collection for conservation; for dissemination on request; for viability testing; for safety duplication; and for repatriation or restoration upon request from the country of seed origin. All the newly harvested accessions are processed and stored in packets for these five different purposes.

Though there are separate medium-term and long-term storage rooms, the GRU plans to consolidate them to reduce costs, but the process of transferring samples from medium-term to long-term storage is necessarily gradual. Hence our cost estimates are based on the current dual storage arrangements (Table 7.5).

Germplasm-health testing

New accessions introduced at CIAT are first tested for disease by the Colombian quarantine unit located next to the GRU. Disseminated materials are tested by the Germplasm Health Laboratory (GHL). The testing is now performed immediately after regeneration and before storage, rather than prior to distribution. About 4000 bean and 1500 forage accessions are tested annually for two types of bacteria, three viruses and a fungus. This number includes both newly regenerated and existing stored samples.

Table 7.5. Annual operating costs (US$, 2000 prices) for bean and forage at the CIAT genebank.

Cost category	Bean				Forage			
	Labour	Non-labour	Subtotal	Capital	Labour	Non-labour	Subtotal	Capital
Medium-term storage	**3,315**	**2,344**	**5,659**	**11,145**	**1,951**	**1,380**	**3,331**	**6,557**
Storage management	1,813	–	1,813	–	1,067	–	1,067	–
Temperature control	902	1,920	2,822	–	531	1,130	1,661	–
Overheads	600	424	1,024	–	353	250	603	–
(Number of accessions)			(31,880)				(18,180)	
Long-term storage	**2,217**	**1,795**	**4,012**	**7,700**	**1,291**	**1,044**	**2,335**	**4,481**
Storage management	910	–	910	–	530	–	530	–
Temperature control	906	1,470	2,376	–	527	855	1,382	–
Overheads	401	325	726	–	234	189	423	–
(Number of accessions)			(12,700)				(7,390)	
Viability testing	**12,137**	**159**	**12,296**	**864**	**6,447**	**2,347**	**8,794**	**3,139**
Viability testing	9,940	130	10,070	–	5,280	1,922	7,202	–
Overheads	2,197	29	2,226	–	1,167	425	1,592	–
(Number of accessions)			(5,000)				(1,000)	
Regeneration	**123,355**	**38,860**	**162,215**	**31,859**	**92,566**	**33,923**	**126,489**	**19,198**
Field operation	72,560	21,389	93,949	–	54,840	21,465	76,305	–
Seed processing	28,468	10,437	38,905	–	20,972	6,318	27,290	–
Overheads	22,327	7,034	29,361	–	16,754	6,140	22,894	–
(Number of accessions)			(5,220)				(3,160)	
Dissemination	**62,778**	**9,383**	**72,161**	**8,673**	**7,681**	**1,236**	**8,917**	**8,673**
Management	5,132	–	5,132	–	628	–	628	–
Seed-health testing	45,390	7,055	52,445	8,673	5,554	863	6,417	8,673
Packing/shipping	893	630	1,523	–	109	149	258	–
Overheads	11,363	1,698	13,061	–	1,390	224	1,614	–
(Number of accessions)			(4,250)				(520)	
Duplication	**559**	**510**	**1,069**	**–**	**–**	**–**	**–**	**–**
Packing/shipping	458	418	876	–	–	–	–	–
Overheads	101	92	193	–	–	–	–	–
(Number of accessions)			(2,180)					
General management	**43,687**	**12,626**	**56,313**	**7,195**	**43,687**	**12,626**	**56,313**	**7,195**
Managerial staff	35,780	–	35,780	–	35,780	–	35,780	–
Other expenses	–	10,341	10,341	–	–	10,341	10,341	–
Overheads	7,907	2,285	10,192	–	7,907	2,285	10,192	–
Total operating cost	**248,048**	**65,677**	**313,725**	**67,436**	**153,623**	**52,556**	**206,179**	**49,243**

Note: See Appendix B for further details.

The focus of the GHL has changed dramatically over the past few years. Until the mid-1990s, most of its operation was geared to testing bean and forage materials for the breeding programmes. Recently, however, germplasm exchange from the breeding programme has virtually halted, and over 90% of the health laboratory's operations were geared to the genebank. Research to establish new methods to test for the presence of key diseases in tropical pastures is integral to their work, and substantial resources have recently been spent for cassava indexation.

Viability testing

In addition to testing newly regenerated materials, delicate species (such as *Phaseolus coccineus, Arachis pintoi* and several grasses) are tested every 5 years for 20 years, hardier species are tested every 5 years for 30 years (except years 10 and 25) and the most durable species, such as woody-seeded legumes, are tested every 10 years from an initial testing in year 5. Once sufficient data have been accumulated, these protocols will be simplified.

Bean accessions are tested in sand-beds for cost efficiency; forages are tested using the conventional paper-rolling method accompanied by a tetrazolium test. Though only 3000 accessions were tested for viability in 2000, the cost elements in Table 7.5 are based on annual averages provided by the genebank manager (5000 for bean and 1000 for forage). As of 2000, 13,000 bean and 6600 forage accessions had undergone testing at CIAT.

Regeneration

Regeneration is conducted in several locations. Most field operations at Palmira (CIAT headquarters) are undertaken by CIAT's field unit on a charge-back basis. However, the genebank uses its own labour and other resources to regenerate genetic material at Palmira along with field activities in other locations, such as Popayan, Quilichao and Tenerife.[9] A total of 8380 accessions of beans and forages (5220 beans and 3160 forages) were regenerated during 2000. Material is usually processed and packed the year after it is harvested to allow for the lengthy drying period. For the purposes of costing, however, we assumed that all the regenerated accessions were cleaned, dried and packed in the same year.

Dissemination and safety duplication

A total of 6949 accessions were disseminated to users around the world in 2000. The GRU ships requested material every 2–3 weeks, while backup duplicates are shipped every 2–3 years. In 2000, 2180 bean accessions were prepared for safety duplication to Brazil and Costa Rica, and held for later shipment. Limited duplication activities occurred for beans and forages in 2000, so the data in Table 7.5 are estimates based on averages over several prior years. Table 7.6 shows the number of disseminated samples in the years leading up to and including 2000.

Economic Analysis

Representative annual costs of conservation

The total annual cost of conserving and distributing cassava at CIAT is estimated at $295,207, of which 74% consists of the cost of *in vitro* conservation and the rest is equally divided between cryoconservation and the field genebank (Table 7.7). In addition, the total annual cost of conserving and distributing bean and forage germplasm at CIAT is $636,583. Together with the annual costs of maintaining cassava germplasm, the total cost of maintaining the current collections of CIAT germplasm per year is estimated at

Table 7.6. Dissemination of germplasm from the CIAT genebank, 1994–2000.

Crop/recipient	1994	1995	1996	1997	1998	1999	2000
Cassava	**550**	**527**	**149**	**219**	**366**	**460**	**2,176**
CIAT/CG	80	192	5	138	278	422	1,936
NARS	470	321	82	46	57		201
Private		9	3	7	7		10
Others		5	59	28	24	38	29
Bean	**8,877**	**7,565**	**8,705**	**10,481**	**8,493**	**9,600**	**4,256**
CIAT/CG	6,007	7,383	6,355	6,339	2,931	8,327	1,468
NARS	2,860	76	1,334	2,727	3,653		2,275
Private	10			4	9		11
Others		106	1,016	1,411	1,900	1,273	502
Forage	**3,231**	**1,133**	**1,320**	**1,053**	**517**	**525**	**517**
CIAT/CG	2,018	632	742	216	200	235	136
NARS	538	312	445	349	121		73
Private	28	89	28	3	20		
Others	647	100	105	485	176	290	308
Total	**12,658**	**9,225**	**10,174**	**11,753**	**9,376**	**10,585**	**6,949**

NARS, National Agricultural Research System.

Table 7.7. Annual total (US$, 2000 prices) and average (US$ per accession, 2000 prices) costs of each operation at the CIAT genebank.

Cost category	Number of accessions	Total capital cost	Total quasi-fixed cost[a]	Total variable cost	Average capital cost	Average quasi-fixed cost	Average variable cost
Cassava		**32,874**	**70,448**	**191,885**	**17.60**	**27.86**	**110.10**
In vitro conservation		**29,341**	**49,300**	**139,888**	**14.74**	**15.72**	**74.76**
Storage	6,080	3,282	6,160	9,358	0.54	1.01	1.54
Subculturing	4,290	11,809	29,931	47,801	2.75	6.98	11.14
Disease indexation	1,200	13,101	4,403	64,323	10.92	3.67	53.60
Dissemination	2,170	1,149	8,806	18,406	0.53	4.06	8.48
Cryoconservation		**3,019**	**9,674**	**25,918**	**2.76**	**9.95**	**30.35**
Storage	2,000	744	1,218	501	0.37	0.61	0.25
Viability testing	300	51	382	2,006	0.17	1.27	6.69
Cryo-operation	1,000	2,224	8,074	23,411	2.22	8.07	23.41
Field genebank		**514**	**11,474**	**26,079**	**0.10**	**2.19**	**4.99**
Field maintenance	5,230	514	11,474	26,079	0.10	2.19	4.99
Bean		**67,436**	**80,688**	**233,039**	**10.61**	**15.65**	**46.39**
Medium-term storage	31,880	12,224	7,599	6,509	0.38	0.24	0.20
Long-term storage	12,700	8,060	2,906	3,922	0.63	0.23	0.31
Viability testing	5,000	1,584	5,225	12,701	0.32	1.05	2.54
Regeneration	5,220	34,737	35,468	149,273	6.65	6.79	28.60
Dissemination	4,250	10,471	27,695	58,544	2.46	6.52	13.78
Duplication	2,180	360	1,795	2,090	0.17	0.82	0.96
Forage		**49,242**	**53,412**	**152,766**	**32.76**	**35.34**	**72.69**
Medium-term storage	18,180	7,636	6,687	5,090	0.42	0.37	0.28
Long-term storage	7,390	4,841	2,442	2,709	0.66	0.33	0.37
Viability testing	1,000	3,858	3,712	10,713	3.86	3.71	10.71
Regeneration	3,160	22,076	29,306	122,524	6.99	9.27	38.77
Dissemination	520	10,831	11,265	11,730	20.83	21.66	22.56
Total cost		**149,552**	**204,548**	**577,690**			

Note: As of 2000, 6080 accessions of cassava (5730 FAO-designated and 290 non-designated accessions) were conserved. For *in vitro* conservation, genebank management costs are allocated as follows for the ICRISAT genebank: storage (10%), subculturing (60%), disease indexation (10%) and dissemination (20%); for cryoconservation, genebank management costs are allocated as follows: storage (10%), viability testing (10%) and operation (80%). For bean and forage, management costs are allocated according to the following percentages: medium-term storage (15%), long-term storage (5%), viability testing (10%), regeneration (40%), duplication (5%) and dissemination (25%). There was no duplication for forages during the sample period, so the cost for beans was used as a proxy.
[a]Total quasi-fixed costs include the costs of senior scientific and technical staff.

$931,790 (Table 7.7). Figure 7.1 shows that almost two-thirds of this annual expense involves variable costs, such as labour and non-labour (62%), while less than a quarter of the total cost is related to capital (16%).

Economic costs

Annual average costs

Annual average costs of conserving cassava vary considerably according to the method used to conserve the material. The cost of conserving an existing cassava accession *in vitro*, including storage and annual subculturing, is $10.34, and the corresponding cost for a new accession is as high as $67.61 due to the cost involved in health testing. We estimate the cost of distributing an *in vitro* sample to be $22.88 per accession, which includes the costs of subculturing, packing and shipping. The with- and without-regeneration cost differentials are more significant for cryoconservation: it costs only $0.86 to conserve an accession for a year if it does not require regeneration or viability testing, while the cost jumps to $40.31 if the sample requires regeneration. The cost of conserving accessions in the field genebank at CIAT is estimated as $7.18 per accession per year, making this the cheapest option for conserving cassava germplasm in the short run if regeneration is required (Table 7.8).[10]

The average costs of conserving seed samples at the CIAT genebank for 1 more year are just 54 cents for beans and 70 cents for forages; if regeneration is needed, the costs increase to $20.03 for beans and to $31.93 for forages; and if the accession is newly introduced in the year in question it costs

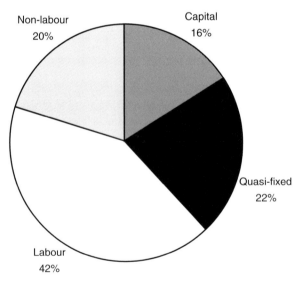

Fig. 7.1. Representative annual costs of maintaining the CIAT genebank holdings. Share of conservation and distribution costs by cost category (US$931,790, 2000 prices).

Table 7.8. Average costs (US$ per accession, 2000 prices) of conserving and distributing a cassava accession for 1 year at the CIAT genebank.

	In vitro		Cryoconservation		
Cost category	Existing accessions	New accessions	Without regeneration	With regeneration	Field genebank
Conservation					
Storage[a]	1.28	1.28	0.86	0.86	7.18
Subculturing[b]	9.06	9.06			
Viability testing				7.96	
Regeneration (cryoconservation)				31.49	
Disease indexation		57.27			
Conservation cost					
In vitro	**10.34**	**67.61**			
Cryoconservation			**0.86**	**40.31**	
Field genebank					**7.18**
Distribution					
Storage	1.28	1.28			
Subculturing	9.06	9.06			
Dissemination	12.54	12.54			
Distribution cost	**22.88**	**22.88**			

[a]Storage costs for the field genebank are the same as the cost of field maintenance.
[b]Storage and subculturing costs for *in vitro* are allocated equally between conservation and distribution.

$25.40 for beans and $48.14 for forages. Similarly, the cost of distributing an accession is $20.73–$40.22 for beans and $44.87–$76.10 for forages depending on whether regeneration is required (Table 7.9).

Average costs in the long run

Under the same baseline assumptions as those used in previous chapters, the average cost of conserving an existing cassava accession in perpetuity is $268.73; for a new accession it is $326.00. A substantial portion of these costs (87% and 72%, respectively) is the cost of subculturing. This suggests that the slow-growth method mentioned earlier (currently being researched to reduce the time required for subculturing) could save considerable conservation costs. The in-perpetuity cost of distributing an accession from the *in vitro* collection is estimated at $307.38, assuming it is disseminated once every 10 years, the contemporary average rate (Table 7.10).

The cost structure of cryoconservation is very different from that of *in vitro* conservation. *In vitro* conservation requires a relatively modest, but frequent, cost of subculturing to maintain the germplasm's viability ($9.06 per accession), while cryoconservation costs $31.49 per accession for regeneration. However, germplasm stored using cryoconservation methods is infrequently regenerated – perhaps only once in every 100 years – whereas subcultured material is regenerated every year, or every other year if the slow-growth method is used. Thus, the in-perpetuity cost of regeneration is only $32.12 under cryoconservation (with 4% interest rate), compared

Table 7.9. Average costs (US$ per accession, 2000 prices) of conserving and distributing a bean or forage accession for 1 year at the CIAT genebank.

Cost category	Existing accessions		New accessions (with regeneration)
	Without regeneration	With regeneration	
Conservation cost			
Bean	**0.54**	**20.03**	**25.40**
Long-term storage	0.54	0.54	0.54
New introduction			
Initial viability testing			3.59
Initial duplication			1.78
Viability testing		1.79	1.79
Regeneration		17.70	17.70
Forage	**0.70**	**31.93**	**48.14**
Long-term storage	0.70	0.70	0.70
New introduction			
Initial viability testing			14.43
Initial duplication			1.78
Viability testing		7.21	7.21
Regeneration		24.02	24.02
Distribution cost			
Bean	**20.73**	**40.22**	
Medium-term storage	0.44	0.44	
Dissemination	20.29	20.29	
Viability testing		1.79	
Regeneration		17.70	
Forage	**44.87**	**76.10**	
Medium-term storage	0.65	0.65	
Dissemination	44.22	44.22	
Viability testing		7.21	
Regeneration		24.02	

with $235.55 for *in vitro* conservation (Table 7.10). Assuming that the viability of germplasm is tested every 20 years, the estimated in-perpetuity cost of cryoconservation increases to $69.11, still substantially lower than the corresponding $268.73 cost of the annually subcultured material.[11] The in-perpetuity, present-value cost of the field genebank is relatively high, at $186.69 per accession, because of recurrent annual planting and harvesting.

As shown above, the average cost of cryoconservation is comparatively high in the short run, but it is the most cost-efficient conservation option in the long run. It should be noted, however, that each conservation method has its own technical and economic pros and cons. Furthermore, conserving material *in vitro* is a prerequisite for storing samples in a cryoconservation environment; thus *in vitro* capabilities must be maintained for newly introduced material even if the option to cryoconserve material is chosen.

Table 7.10. Present values (US$ per accession, 2000 prices) of conserving and distributing a cassava accession in perpetuity at the CIAT genebank.

Cost category	In vitro			Cryoconservation			Field genebank		
	2%	4%	6%	2%	4%	6%	2%	4%	6%
Conservation									
Storage	65.08	33.18	22.54	43.84	22.35	15.19	366.20	186.69	126.85
Subculturing[a]	462.05	235.55	160.06						
Viability testing[b]				24.34	14.64	11.57			
Regeneration[c] (cryoconservation)				36.53	32.12	31.58			
Disease indexation	57.27	57.27	57.27						
Conservation cost									
In vitro									
Existing accession	**527.13**	**268.73**	**182.60**						
New accession	**584.40**	**326.00**	**239.87**						
Cryoconservation				**104.71**	**69.11**	**58.34**			
Field genebank							**366.20**	**186.69**	**126.85**
Distribution									
Storage	65.08	33.18	22.54						
Subculturing	462.05	235.55	160.06						
Dissemination[d]	69.80	38.65	28.40						
Distribution cost	**596.93**	**307.38**	**211.00**						

Note: Some figures in this table were calculated using equations (1) and (2) in Appendix A.
[a]From the time of acquisition, subculturing for *in vitro* occurs every year.
[b]Viability testing for cryoconservation occurs every 20 years.
[c]Regeneration for cryoconservation occurs every 100 years.
[d]Dissemination of an accession occurs every 10 years.

For beans and forages, viability testing is assumed to take place every 10 years for the base collection and every 5 years for the active collection; the regeneration interval is 50 and 25 years, respectively; and an accession is assumed to be disseminated once every 10 years. Using the baseline 4% interest rate, the average cost of conserving an existing bean accession in perpetuity is $20.90; for forages it is $37.36; for a newly acquired accession,

Table 7.11. Present values (US$ per accession, 2000 prices) of conserving and distributing a bean or forage accession in perpetuity at the CIAT genebank.

Cost category	Existing accession			New accession		
	2%	4%	6%	2%	4%	6%
Conservation						
Beans						
Long-term storage	27.42	13.98	9.50	27.42	13.98	9.50
New introduction						
Initial viability testing				3.59	3.59	3.59
Initial duplication				1.78	1.78	1.78
Viability testing	8.19	3.73	2.27	8.19	3.73	2.27
Safety duplication	1.05	0.29	0.10	1.05	0.29	0.10
Regeneration/viability testing (50 years)[a]	10.46	2.90	1.02	28.16	20.59	18.71
Conservation cost	**47.12**	**20.90**	**12.89**	**70.19**	**43.96**	**35.95**
Forages						
Long-term storage	35.54	18.12	12.31	35.54	18.12	12.31
New introduction						
Initial viability testing				14.43	14.43	14.43
Initial duplication				1.78	1.78	1.78
Viability testing	32.94	15.02	9.12	32.94	15.02	9.12
Safety duplication	1.05	0.29	0.10	1.05	0.29	0.10
Regeneration/viability testing (50 years)	14.20	3.93	1.38	14.20	3.93	1.38
Conservation cost	**83.73**	**37.36**	**22.91**	**99.94**	**53.57**	**39.12**
Distribution						
Bean						
Medium-term storage	22.57	11.51	7.82			
Viability testing	17.22	8.27	5.30			
Dissemination[b]	112.95	62.54	45.95			
Regeneration/viability testing (25 years)	17.16	7.72	4.36			
Distribution cost	**169.90**	**90.04**	**63.43**			
Forage						
Medium-term storage	33.04	16.84	11.44			
Viability testing	69.30	33.29	21.33			
Dissemination	246.15	136.30	100.14			
Regeneration/viability testing (25 years)	23.30	10.49	5.92			
Distribution cost	**371.79**	**196.92**	**138.83**			

Note: Some figures in this table were calculated using equations (1) and (2) in Appendix A.
[a]Regeneration costs include germination testing after regeneration; regeneration for distribution commences at the 25th year, and then occurs every 50 years.
[b]Dissemination is assumed to occur every 10 years.

the cost ranges from $43.96 for beans to $53.57 for forages. The costs of distributing an accession – $90.04 for beans and $196.92 for forages – are much higher, given the higher costs of maintaining the active collection and the additional costs of disseminating samples (Table 7.11).

Total costs in the long run

Table 7.12 presents the total costs of maintaining the current collection at the CIAT genebank. If all the cassava accessions (6080 accessions) are conserved *in vitro*, the total conservation cost in perpetuity is $2,366,887, including the capital cost. The corresponding figures for maintaining the complete cassava collection using cryoconservation methods are $655,669, and $1,150,596 for material kept in a field genebank. Taking into account the cost of distributing samples from *in vitro* ($1,898,776) and presuming all the accessions were conserved using all three techniques yields an in-perpetuity cost of $6,071,928 for cassava.[12]

The total cost of conserving and distributing the current collection of bean and forage germplasm in perpetuity is estimated at $3,756,450 for beans and $4,857,797 for forages. Labour and operating costs (designated as non-capital cost in Table 7.12) to conserve the entire holdings of beans and forages in perpetuity amount to $7,687,771, while the corresponding capital cost accounts for only $926,476 (Table 7.12).

Notes

[1]Plants conserved *in vitro* can be maintained free from disease contamination, enabling safe dissemination.

[2]The accession numbers used for our cost estimates were developed in November 2000 and represent estimates of the size of the entire CIAT holdings as of mid-2000. The data in Table 7.1 were compiled in April 2003 and include only FAO-designated cassava and forage material.

[3]All collections are conserved under medium-term conditions. As of 2000, about 40% of the total collection was conserved in the long-term storage room. A few forage species (fewer than 20 accessions) are also maintained in a field genebank because they do not produce enough seed for storage.

[4]In general, conservation of cassava as seeds would allow the conservation of specific genes but not necessarily the range of genotypes found within a given species. For wild species of *Manihot*, however, a cost-effective and successful method is to conserve well-dried botanical seeds in long-term storage because the alleles present in a source population can best be preserved in this way (though sample frequencies may vary in comparison with the original populations).

[5]The tests are the enzyme-linked immunosorbent assay (ELISA) method for cassava common mosaic virus (CCMV) and cassava X virus (CsXV), the grafting method for frog skin disease (FSD) and the polymerase chain reaction (PCR) method for cassava vein mosaic virus (CVMV).

[6]Only a few can be recovered from each plantlet (usually three or four), so several plantlets are needed.

[7]Many research units at CIAT use liquid nitrogen, and there was a proposal to purchase liquid nitrogen-producing equipment and generate it in-house. This equipment

Table 7.12. Total costs of conservation and distribution in perpetuity at the CIAT genebank.

Crop	Number of accessions	Per-accession cost (US$ per accession, 2000 prices)			Total cost (US$, 2000 prices)		
		Conservation	Distribution	Total	Conservation	Distribution	Total
Cassava	6,080	**524.53**	**307.38**	**831.91**	**4,173,152**	**1,898,776**	**6,071,928**
In vitro		**268.73**	**307.38**	**576.11**	**2,366,887**	**1,898,776**	**4,265,663**
Non-capital		268.73	307.38	576.11	1,633,890	1,868,896	3,502,786
Capital		–	–	–	732,997	29,880	762,877
Cryoconservation		**69.11**	–	**69.11**	**655,669**	–	**655,669**
Non-capital		69.11	–	69.11	420,208	–	420,208
Capital		–	–	–	235,461	–	235,461
Field genebank		**186.69**	–	**186.69**	**1,150,596**	–	**1,150,596**
Non-capital		186.69	–	186.69	1,135,063	–	1,135,063
Capital		–	–	–	15,533	–	15,533
Bean	30,900	**25.19**	**96.38**	**121.57**	**778,260**	**2,978,190**	**3,756,450**
Non-capital		20.90	90.05	110.95	645,842	2,782,500	3,428,342
Capital		4.29	6.33	10.62	132,418	195,690	328,108
Forage	18,180	**43.61**	**223.60**	**267.21**	**792,791**	**4,065,006**	**4,857,797**
Non-capital		37.37	196.93	234.30	679,307	3,580,122	4,259,429
Capital		6.24	26.67	32.91	113,484	484,884	598,368
All crops	**55,160**	**593.33**	**627.36**	**1,220.69**	**5,744,203**	**8,941,972**	**14,686,175**

would reduce the cost of liquid nitrogen dramatically. Currently, the cost includes transportation from the local vendor and insurance.

[8]Now, all the accessions are planted in April and harvested in February of the following year.

[9]The site at Tenerife, which has been used to regenerate about 4000 bean accessions annually, was abandoned in mid-2000 for security reasons.

[10]CIAT maintains its field genebank for conservation and other (e.g. crop breeding and characterization) purposes. Consequently our per-accession estimates of the respective conservation costs are likely to be lower than if the field facilities were operated solely for conservation purposes (i.e. they reflect some savings because of economies of scale and scope).

[11]Because cryoconservation is still experimental, CIAT initially plans to test the viability of cryoconserved accessions every 5 years, but with an expectation that the viability testing interval will lengthen to 20 years.

[12]Another option is to conserve the cassava collection *in vitro* for medium-term storage and cryoconserve the material designated for long-term storage. Using this protocol, the total cost of conserving CIAT's current cassava holdings in perpetuity would be $4,921,333.

8 Policy and Management Implications

BONWOO KOO, PHILIP G. PARDEY AND BRIAN D. WRIGHT

Precisely how much does it cost to conserve the entire CG holdings into the far distant future? What are the cost implications of changing conservation protocols or consolidating genebank facilities? What are the revenue consequences of providing some genebank operations on a fee-for-service basis? In this chapter we lay out the economic implications of a number of important policy and management options, drawing on the cost data discussed in the previous chapters. These data were compiled over a period of several years and, consequently, at different pricing levels (1996 for CIMMYT, 1998 for ICARDA, 1999 for IRRI and ICRISAT and 2000 for CIAT, as noted in Chapter 2). For the purposes of this chapter, we converted all costs per accession to 2000 prices, using a weighted average of the producer price index for the G7 countries, constructed from data reported in OECD (2000) and World Bank (2000).[1] In addition, the total costs were calculated using the estimates of the total number of accessions of each centre in 2001 (Table 1.1). Accordingly, the costs reported in this chapter vary a little from the corresponding costs identified in previous chapters.[2]

Comparison of Operating Costs

Table 8.1 presents a breakdown of the baseline, per-accession costs per year (inclusive of annualized capital costs) for each centre's operation and its respective crops. For most crops at most centres, the differences between medium- and long-term storage costs are much smaller than the differences in regeneration costs among crops. As pointed out in the individual chapters, the general pattern shows that cross-pollinating species (like maize and pigeonpea) and wild species (like wild groundnut or wild rice) are much more costly to regenerate than other crop types. There are also significant locational or institutional differences relating to regenerating crops,

Table 8.1. Operational costs per accession (US$ per accession, 2000 prices) for each crop in selected genebanks.

Centre	Crop	Acquisitions	Medium-term storage	Long-term storage	Viability testing	Regeneration	Characterization	Duplication	Dissemination
CIAT	Cassava								
	In vitro[a]	68.19	3.09	1.23	8.13	20.87			13.07
	Cryoconservation			7.28		33.70			
	Field genebank[b]								
	Common beans		0.44	0.92	4.22	35.71		4.24	26.95
	Forages		0.65	1.12	15.08	51.91		4.24	51.21
CIMMYT	Wheat	3.30	0.37	0.48	1.36	6.63		0.44	4.20
	Maize	9.47	3.04	2.16	4.79	221.02		5.53	35.45
ICARDA	Cereals	6.10	0.55	0.47	2.70	10.09	1.55	2.51	3.71
	Forages	6.10	0.55	0.47	2.70	12.78	1.46	2.51	3.71
	Chickpea	6.10	0.55	0.47	2.70	13.88	2.00	2.51	3.71
	Lentil	6.10	0.55	0.47	2.70	17.85	2.22	2.51	3.71
	Faba bean	6.10	0.55	0.47	2.70	17.59	2.65	2.51	3.71
ICRISAT	Sorghum	5.27	1.51	1.32	1.26	19.88	11.10	4.39	2.58
	Pearl millet	5.27	1.51	1.32	1.26	51.05	12.67	4.39	2.58
	Pigeonpea	5.27	1.51	1.32	1.26	60.31	6.38	4.39	2.58
	Chickpea	5.27	1.51	1.32	1.26	21.51	5.11	4.39	2.58
	Groundnut	5.27	1.51	1.32	1.26	28.18	18.34	4.39	2.58
	Wild groundnut	5.27	1.51	1.32	1.26	249.00	26.36	4.39	2.58
IRRI	Cultivated rice	6.51	0.87	0.47	1.54	33.90	10.15	1.74	9.75
	Wild rice	6.51	0.87	0.47	1.54	114.74	7.62	1.74	9.75

[a]The acquisition costs for material to be held in vitro represent the costs of screening the health of the sample by disease-indexing methods. Regeneration costs for material held in vitro represent the costs of subculturing the accession. Most cassava is distributed in the form of in vitro samples.
[b]As a practical matter, conserving cassava in a field genebank is more properly thought of as a medium-term undertaking, but we included it here under long-term storage to reflect its conservation intent.

such as wheat at CIMMYT versus ICARDA, forages at ICARDA versus CIAT and chickpea at ICRISAT versus ICARDA.[3]

Figure 8.1 compares the costs of conserving an accession for 1 year (panel a) with the present-value costs of conserving an accession in perpetuity (panel b). If regeneration is not required, holding over a seed sample for 1 year costs less than $1.50 – with the exception of maize, which costs $2.16 per accession, and cassava when conserved *in vitro*, which costs $11.98 per accession. Electricity and the annualized capital cost of the storage facility are the primary elements of these storage costs, along with a small amount for the maintenance of storage equipment. Storage costs at ICARDA are comparatively low as a result of cheaper labour and electricity, in contrast to ICRISAT, where electricity is expensive. The comparatively high cost of storing maize stems from its large seed size and hence its need for greater storage space and larger, more expensive storage containers.

Considering storage costs in perpetuity – thereby accounting for recurring viability testing and regeneration – the cost hierarchy shifts considerably. As we have already seen, wild and weedy species and cross-pollinating crops that are expensive to regenerate become relatively more expensive in present-value terms when costs are accumulated over the long term, so CIAT's forages and IRRI's wild rice become more expensive to conserve than ICRISAT's chickpea and sorghum.

Table 8.2 provides a representative snapshot of the total annual conservation and distribution costs incurred by each of the CG centres. The estimates for CIAT, CIMMYT, ICARDA, ICRISAT and IRRI, which together account for 87% of the entire CG germplasm accessions, were used as the basis for estimating the costs for the remaining six CG centres with active conservation and distribution programmes (as elaborated below). These total costs include all the labour and operational costs incurred to provide core conservation and distribution services for 1 year, as well as an estimate of the annualized cost of the recurrent capital expenditures required to build and equip the genebanks. Based on the assumptions that underlie these estimates, the total annual cost for all 11 CG genebanks is $5.7 million. Table 8.2 illustrates that accession numbers per centre are not particularly good comparative indicators of conservation costs. As noted in Chapter 1, many other factors affect these costs – some intrinsic to the crop in question, others relating to the location or the institutional arrangements in which a genebank operates.

Operational Issues of Genebank Management

Cassava storage at CIAT

In Chapter 7, we show that subculturing of cassava germplasm is very costly in the long run because of the frequency with which it must be undertaken. The slow-growth method developed by CIAT extends the period between successive rounds of subculturing from 12 to approximately 24 months, thus reducing the material and labour costs associated with the process and

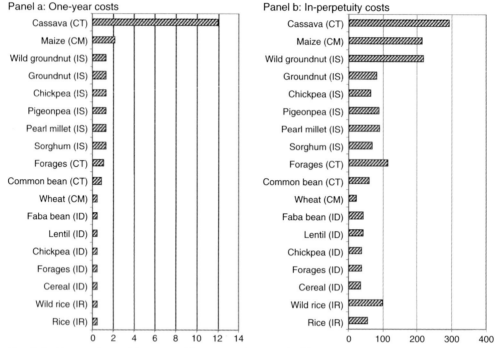

Fig. 8.1. Comparison of 1-year and in-perpetuity costs (US$ per accession, 2000 prices). CT, CIAT; IS, ICRISAT; ID, ICARDA; CM, CIMMYT; IR, IRRI.

Table 8.2. Representative total annual costs (US$, 2000 prices) of conservation and distribution in CG genebanks.

Centre	Number of accessions	Conservation cost	Distribution cost	Total cost
CIAT	63,644	354,159	573,444	927,603
CIMMYT	179,998	350,891	686,734	1,037,625
ICARDA	122,467	179,286	261,259	440,545
ICRISAT	113,501	347,224	419,244	766,468
IRRI	99,132	217,411	403,142	620,553
Other	87,338	913,879	1,020,139	1,934,018
CG total	**666,080**	**2,362,850**	**3,363,962**	**5,726,812**

Note: These estimates were calculated at a 4% baseline interest rate using equation (3) in Appendix A. Material at 'other' centres mainly consists of vegetatively propagated species; we used only CIAT's cassava cost to extrapolate the costs of this material. Thus, some crop-specific characteristics of the vegetatively propagated species may not have been considered.

enabling the genebank to handle larger numbers of accessions more cost-effectively. While material conserved *in vitro* and clonally propagated is spared the genetic drift inherent in seed samples that are regenerated sexually, there is still a risk of somaclonal variation (mutation of the cell lineages induced by the growth medium and experimental conditions), the extent of which is likely to be reduced by switching from a conventional to a slow-growth medium.

Table 8.3 compares the in-perpetuity costs of conserving cassava germplasm under different methods – the current method with a 12-month cycle of subculturing versus the slow-growth method involving a 24-month cycle. While the annual storage cost is the same for both methods, the in-perpetuity cost of subculturing is reduced by almost one-half. Table 8.3 shows that the long-term estimated saving in switching to this new method is $230.92 per accession. With a total of 6080 accessions of cassava conserved *in vitro* at CIAT, the total cost saving from adopting the slow-growth method is $1,403,994. Consequently, if the method can be developed for anything less than $1.4 million, the research is a worthwhile investment, even taking a narrow, CIAT-centric, perspective.[4]

Groundnut storage at ICRISAT

The ICRISAT genebank has traditionally stored groundnut germplasm as pods, with their shells intact ('in-shell storage'). But, like maize, the large size of unshelled pods necessitates storage in 5 litre plastic jars for each accession, with the result that the 15,342-accession collection occupies two large storage modules (76.7 m² each). To economize on space and costs, the genebank has begun to shell the accessions and store only the shelled groundnut seed ('shelled storage'). In this way, only one medium-sized storage module is needed to store the complete collection.

Using the cost information in Chapter 5, Table 8.4 compares the costs of storing groundnut as in-shell type versus shelled type. Considering the nature of the cold-storage operation, the amount of quasi-fixed labour required is insensitive to the method used.[5] However, the two-module storage of in-shell groundnut consumes about 240,000 kWh of electricity per year, while the single-module storage of shelled groundnut uses only 80,000 kWh per year. Thus, variable costs of electricity are roughly halved as a result of the change. Consequently, the average storage cost is estimated at $2.07 per accession for in-shell storage and $0.77 per accession for shelled storage, producing a cost saving of $1.30 per accession or about $20,000 for the full collection (15,340 accessions).

Table 8.3. *In vitro* costs (US$, 2000 prices) under different subculturing methods at the CIAT genebank.

	In-perpetuity cost per accession			Total in-perpetuity costs (6080 accessions)		
Activity	Current method (12 months)	New method (24 months)	Cost saving	Current method (12 months)	New method (24 months)	Cost saving
Storage	66.36	66.36	–	403,469	403,469	–
Subculturing	471.10	240.18	230.92	2,864,288	1,460,294	1,403,994
Total	**537.46**	**306.54**	**230.92**	**3,267,757**	**1,863,763**	**1,403,994**

Note: Costs exclude capital costs for both conservation and distribution purposes.

Table 8.4. Costs (US$, 2000 prices) of different storage methods for groundnut at the ICRISAT genebank.

Cost category	Current protocol: in-shell storage	Proposed protocol: shelled storage
Storage		
Quasi-fixed cost	1,500	1,500
Variable cost		
Electricity	28,800	9,600
Equipment maintenance (non-labour)	500	250
Equipment maintenance (labour)	900	450
Total cost	31,700	11,800
Average cost	**2.07**	**0.77**
Regeneration		
Average cost[a]	**25.07**	**26.32**
Regeneration interval	11 years	10 years
In-perpetuity cost[b]		
Storage cost[c]	53.82	20.02
Regeneration cost[d]	71.48	81.06
Total cost per accession	**125.30**	**101.08**
Total cost for all accessions	**1,922,102**	**1,550,567**
(Number of accessions)		(15,340)

[a]The cost for in-shell storage is taken from Table 5.7; the cost for shelled storage is assumed to be 5% higher than that of in-shell storage.
[b]Used 4% rate of interest.
[c]See equation (3) in Appendix A.
[d]See equation (2) in Appendix A.

Shelled storage does involve higher regeneration cost because of the extra labour needed to remove the shells after harvesting. In the absence of specific data, we estimate, conservatively, that the cost of regenerating shelled seeds is about 5% greater than for groundnut stored with their shells. In addition, there is a general consensus that the lifespan of seeds stored without their shells will be shorter, requiring more frequent regeneration, although the experiments done to date indicate that the difference in regeneration intervals is small (less than 3 months). For in-shell storage we estimate an 11-year regeneration interval; for the shelled method, the interval is a conservative 10 years.

Taking all these factors into account, the in-perpetuity cost of conserving a groundnut accession is estimated as $125.30 under in-shell storage and $101.08 under shelled storage (using our baseline, 4% interest rate). The total cost of conserving 15,340 groundnut accessions in perpetuity would be $1,922,102 for in-shell storage and $1,550,567 for shelled storage, netting a cost saving of $371,535. This result, however, is sensitive to location and institutional idiosyncrasies. For example, in India labour (used for shelling) is inexpensive compared with electricity (used for storage), which is not necessarily the case in other locations.

Economies of scale

In addition to the 666,000 accessions held in the 11 genebanks maintained by the CGIAR, there are about 5.6 million accessions stored in more than 1300 genebanks worldwide (FAO, 1998). This leads to the following question: Would consolidating these holdings in fewer facilities provide economic gains? CIMMYT data offer some insights, at least for wheat and maize.

By world standards the CIMMYT holding is large, but not the largest. Around the mid-1990s, the Institute of Crop Germplasm in China had a total of about 300,000 accessions in long-term storage while the National Seed Storage Laboratory in the United States held 268,000 seed samples in its collection (FAO, 1998). As of 1996, the germplasm conserved at CIMMYT occupied 30–40% of the full capacity of the genebank facility – 123,000 of the 390,000 accession capacity for wheat, and 17,000 of the 67,000 accession capacity for maize. We estimated the storage costs of operating the genebank in 1996 with its total of 140,000 wheat and maize accessions to be $655,725 (Table 3.6). If the genebank were operating at full capacity with the cost structure at its 1996 capacity, the annual storage costs would be $1,148,098 ($568,005 for wheat and $580,094 for maize). However, operating at full capacity while realizing capital cost economies – i.e. scaling up the size of the holdings from 140,000 to 457,000 accessions without changing the annualized cost of capital – would involve a total annual cost of $896,466 (an annual saving in storage costs of $251,632). Allowing for capital plus quasi-fixed cost economies would involve a full-capacity cost of $769,424. Spreading the unchanging annualized cost of the building structure and equipment and the senior technical staff and the genebank manager over a larger number of conserved samples would incur an annual cost saving of $378,674. These results illustrate the sizable savings to be made from consolidating collections in one location compared with storing them at separate facilities.

Our calculations imply some considerable economies from centralizing storage of all cultivars of a crop and avoiding excessive duplication of storage facilities. Given the relatively modest cost of black-box or other forms of safety duplication, conservation economics and security imperatives can be jointly satisfied with one central genebank and duplicates held in other parts of the world. One possible scenario would be to set up a central genebank for long-term conservation, with various local genebanks operating active collections. However, the optimal scenario depends on transport and communication costs, on the relative conservation costs of active collections and long-term collections and on the different effects of the environment on different crops – issues for further study. At least one duplicate should be located outside the risk of political embargo, military action or terrorism that could disrupt international access.

Charging for distribution services

Distributing seed samples (mostly on demand) is a major undertaking with significant cost implications for the CG genebanks. Providing these services free of charge is increasingly questionable in the light of dwindling CG budgets. Some centres have charged for some of these services on an occasional, ad hoc basis, but there are at present no clear centre-specific or CG-wide policies or practical guidelines for recovering these costs based on 'user pays' or other principles.[6] One option is to charge all or parts of the distribution (distinct from conservation) costs to all those requesting samples. Other options are to segment the demand for seed shipments, charging on a graduated basis according to the size of the shipment, or differentiating between public and private requests, or invoking ability-to-pay approaches whereby institutes and individuals in poor countries are charged little or nothing and others pay part or all of the costs involved.

Knowing something about the costs involved is a prerequisite for choosing an appropriate cost-recovery strategy. Table 8.5 reports the size of the partial and full costs of distributing each crop from each centre, based on the cost information in Table 8.1. The partial costs include the costs of multiplying as well as shipping and handling the seed samples, while the full costs also include the costs of storing (medium-term) and characterizing the seed. The charges vary greatly by crop type – from as low as $12.56 for wheat accessions at CIMMYT to $280.71 for wild groundnut accessions at ICRISAT.[7]

Table 8.5. Distribution costs per accession (US$ per accession, 2000 prices) for each crop in selected genebanks.

Centre	Crop	Partial charge[a]	Full charge[b]
CIAT	Cassava		
	In vitro	33.94	37.03
	Cryoconservation	41.83	41.83
	Field genebank		
	Common bean	66.88	67.32
	Forages	118.20	118.84
CIMMYT	Wheat	12.19	12.56
	Maize	261.27	264.31
ICARDA	Cereals	16.49	18.59
	Forages	19.18	21.19
	Chickpea	20.28	22.82
	Lentil	24.25	27.02
	Faba bean	23.99	27.20
ICRISAT	Sorghum	23.72	36.32
	Pearl millet	54.90	69.07
	Pigeonpea	64.15	72.04
	Chickpea	25.36	31.97
	Groundnut	32.03	51.87
	Wild groundnut	252.84	280.71
IRRI	Cultivated rice	45.20	56.22
	Wild rice	126.03	134.52

[a]The partial charge includes the costs of regeneration, viability testing and dissemination.
[b]The full charge adds the costs of medium-term storage and characterization to the partial charge.

To be clear, these costs represent the expense of maintaining and disseminating conserved seed, independent of the rights to use the seed. The material transfer agreements that currently define the rights of recipients of in-trust material held in CG genebanks prohibit any form of intellectual property rights being sought over the designated accessions, although rights can be pursued over derivatives based on the shipped samples. Switching to some form of fee-for-service structure is as much about who should pay for such services as about how much should be paid. At present it is taxpayers in rich countries who foot most of the bill via their governments' annual contributions to the CGIAR. However, the long-term sustainability of this form of financing is questionable, leading to efforts to establish a conservation fund and, perhaps, implement some user-fee options for some parts of the genebank operations (like distribution services) for which users can be readily identified and segmented into various groups.

A Conservation Endowment Fund

The spread of modern crop varieties worldwide during the 20th century is seen as a two-edged sword. On the one hand, the use of modern varieties resulted in historically unprecedented increases in crop yields in many parts of the world, generating billions of dollars of benefits to producers (in the form of improved productivity and reduced costs of production) and consumers (in the form of reduced prices for food and fibre) (Alston *et al.*, 2000). On the other hand, the widespread uptake of scientifically bred varieties undercut the use of farmer-developed varieties. And, as the land cultivated under more modern farming methods increased, fringe habitats supporting the wild and weedy ancestors of today's agricultural crops were reduced and exposed to new selection and evolutionary pressures.

The danger of excessive reliance by farmers and breeders on a narrowing genetic base was dramatized in the early 1970s by the infestation of southern corn leaf blight in a large portion of the United States hybrid maize crop. The vulnerability of the maize crop was related to the widespread use of cytoplasm male sterility as a means of reducing the cost of breeding hybrid maize by eliminating the need for mechanically removing the tassels that produce the pollen from the female parent used in the hybridization process. These specific developments, combined with growing biodiversity and environmental concerns generally, catalysed efforts worldwide to greatly expand the amount of agricultural biodiversity conserved in *ex situ* genebanks for use in crop breeding.

Since then, modern biotechnologies that provide new and less costly ways of screening crop samples for useful traits have increased the usefulness of genebanks to breeders. The recent surge in interest in searching for valuable traits in conserved genetic resources has generated concerns about the sustainability of conservation efforts due to the mismatch between the generally short-term nature of the financial support for crop conservation and the

long-term nature and intent of the undertaking. To sustain the long-term conservation and use of agricultural germplasm held in *ex situ* genebanks, one
option is to establish a financial endowment, the annual earnings from which
can underwrite conservation efforts into the far distant future.[8]

The costing evidence in Table 8.1 forms the basis for our estimates of an
endowment fund to conserve the current CG holdings in perpetuity. As
mentioned, to do this we extrapolated annualized, per-accession costs for
the five centres that were directly surveyed to the six centres that were not
directly studied.[9] This per-accession method of extrapolating costs may
underestimate the total conservation costs for smaller genebanks, given the
indivisibilities of some capital equipment and facilities that are required
regardless of the size of the genebank.

Because of the substantial differences in conserving and regenerating
tree species compared with conventional crop species, we relied on annual
budget data and informed estimates from the manager of ICRAF's genetic
resource programme to generate approximate, but representative, estimates
of the annual conservation, multiplication and dissemination costs.
Maintaining the headquarters operation, which includes a medium-term
storage facility and ancillary buildings, along with a wide network of on-
farm conservation and regeneration sites in ten countries around the world,
is estimated to cost a total of about $800,000 annually – 80% of which was
allocated to the genebank functions of ICRAF included in this study. These
costs were split by a ratio of 4:6 between conservation and distribution.

Baseline estimates

Our best estimate is that $149 million, invested at a 4% per annum real
interest rate (or a nominal rate of, say, 7% assuming an average 3% per
annum inflation rate over the long run), would generate a real annual revenue of $5.7 million – sufficient to underwrite the costs of conserving and
distributing the current holdings of all 11 CG genebanks in perpetuity
(Table 8.6). About 20% of the funds (nearly $30 million) would be allocated
to ongoing purchases of equipment and genebank buildings as they needed
replacing. The balance would be set aside to fund recurring non-capital
costs related to conservation and distribution.

Figure 8.2 illustrates the estimated centre-specific shares of the overall
fund. The conservation and distribution activities undertaken by the five
directly costed centres, which collectively conserve 87% of the CG's current
germplasm holdings, could be supported with 66% of the total endowment
fund. The remaining 34% would fund activities at the six remaining centres.
These estimates indicate that 13% of the genebank holdings account for 34%
of the total costs. This reflects the reality that vegetatively propagated material, constituting a large part of the IITA, Centro Internacional de la Papa (CIP)
and INIBAP collections, and the tree species conserved by ICRAF are intrinsically costly to store and regenerate. CIAT and CIMMYT constitute 16% and
18% of the total costs, respectively. These centres are located in comparatively

Table 8.6. The conservation endowment fund (US$, 2000 prices) of CG collection.

Centre	Crop	Number of accessions	Conservation cost	Distribution cost	Total cost
CIAT	Cassava	8,060	4,607,383	2,120,296	
	Common bean	31,400	1,942,532	5,034,754	
	Forages	24,184	2,658,223	7,754,497	
	Total	**63,644**	**9,208,138**	**14,909,547**	**24,117,685**
CIMMYT	Wheat	154,912	3,743,844	5,907,775	
	Maize	25,086	5,379,326	11,947,314	
	Total	**179,998**	**9,123,170**	**17,855,089**	**26,978,259**
CIP	Potato	7,639	3,668,612	2,009,546	
	Sweet potato	7,659	3,678,217	2,014,808	
	Andean roots/tubers	1,495	717,970	393,281	
	Total	**16,793**	**8,064,799**	**4,417,635**	**12,482,434**[a]
ICARDA	Cereal	60,013	2,193,716	3,164,959	
	Forages	30,528	1,160,736	1,707,174	
	Chickpea	11,219	439,778	641,903	
	Lentil	9,962	415,729	616,747	
	Faba bean	10,745	451,478	661,945	
	Total	**122,467**	**4,661,437**	**6,792,728**	**11,454,165**
ICRAF	Agroforestry trees	**10,025**	**7,488,000**	**11,232,000**	**18,720,000**[a]
ICRISAT	Sorghum	36,721	2,591,397	2,854,548	
	Pearl/small-variety millet	30,644	2,766,295	3,510,920	
	Pigeonpea	13,544	1,210,405	1,699,923	
	Chickpea	17,250	1,130,474	1,374,307	
	Groundnut	14,892	1,230,678	1,303,838	
	Wild groundnut	450	98,581	156,801	
	Total	**113,501**	**9,027,830**	**10,900,337**	**19,928,167**
IITA	Bambara groundnut	2,029	167,677	177,645	
	Banana	400	192,099	105,226	
	Cassava	3,529	1,694,794	928,353	
	Cowpea	16,001	1,048,621	1,274,799	
	Soybean	3,053	188,871	489,526	
	Wild *Vigna*	1,684	185,100	539,967	
	Miscellaneous legumes	400	43,967	128,258	
	Yam	3,700	1,776,917	973,337	
	Total	**30,796**	**5,298,046**	**4,617,111**	**9,915,157**[a]
ILRI	Forages	**13,204**	**1,451,339**	**4,233,806**	**5,685,145**[a]
INIBAP	*Musa*	**1,143**	**437,070**	**300,682**	**737,752**[a]
IRRI	Cultivated rice	94,564	5,198,429	9,582,545	
	Wild rice	4,568	454,262	899,158	
	Total	**99,132**	**5,652,691**	**10,481,703**	**16,134,394**
WARDA	Rice	**15,377**	**845,314**	**1,558,212**	**2,403,526**[a]
CG Total		**666,080**	**61,257,834**	**87,298,850**	**148,556,684**

[a]These figures represent indirect estimates formed by extrapolating direct costs from crops and centres.

advanced developing countries in Latin America, where wage rates are high by developing-country standards. They also maintain sizeable holdings of crops that are intrinsically costly to conserve – vegetatively propagated cassava (at CIAT) and cross-pollinating maize (at CIMMYT).

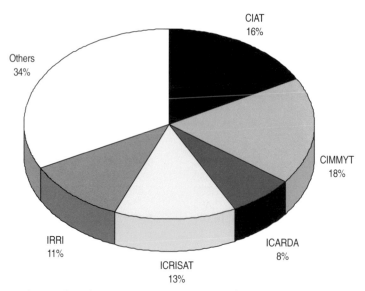

Fig. 8.2. Centre shares of total CG conservation costs. Total = $149 million.

Sensitivity analysis

Our baseline cost estimates build on a number of assumptions. Sensitivity analysis tests the effect of changes in those assumptions on funding estimates, in present-value terms. Because the endowment fund represents the present value of the in-perpetuity costs it is designed to support, significant cost elements that repeat at regular intervals are likely to have a large effect on the estimated size of the endowment fund. Table 8.1 makes it clear that the regeneration costs represent a significant share of the non-capital costs. Thus regenerating material at longer or shorter cycles will lower or raise costs accordingly. The interest rate is also a key component of any present-value calculation; lower rates tend to raise the present-value of future costs.

We tested the sensitivity of our best endowment-fund estimate ($149 million) to changes in these two elements by recalculating the fund figure using sets of interest rates and regeneration cycles given in Table 8.7. In scenario A, the storage lives are comparatively short, requiring more frequent regeneration and viability testing, but the interest rate is reduced to 2%. For scenario C, the interest rate is 6% and the storage lives are much longer, and the cycle of regeneration and viability testing is thus less frequent. Scenario B represents a medium (and seemingly most plausible) regeneration cycle used to form the baseline estimates in Table 8.7. Figure 8.3 shows that with this combination of key assumptions the size of the endowment fund could be as low as $100 million (under scenario C with a high, 6%, rate of interest) or as high as $325 million (under scenario A with a low, 2%, rate of interest).

Table 8.7. Duration of genebank activity cycles under alternative scenarios (from FAO/IPGRI, 1994; Sackville Hamilton and Chorlton,1997; personal communications from CG genebank managers).

	Duration (years)		
Activity cycle	Scenario A	Scenario B	Scenario C
For seed storage			
Long-term storage regeneration cycle	30	50	100
Long-term storage viability-testing cycle	5	10	10
Medium-term storage regeneration cycle	15	25	50
Medium-term storage viability-testing cycle	5	5	10
Dissemination cycle	5	5	5
For *in vitro* conservation			
Subculturing cycle	1	1.5	2
For cryoconservation			
Regeneration cycle	50	100	150
Viability-testing cycle	10	10	15

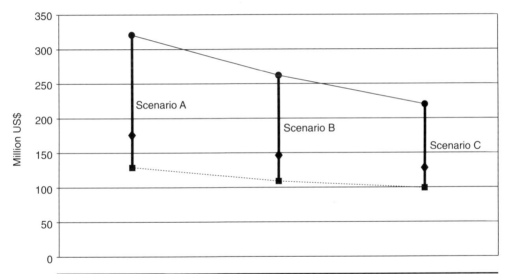

		Scenario and costs (million US$, 2000 prices)		
Table legend	Interest rates	Scenario A: short	Scenario B: medium	Scenario C: long
—●—	2%	325	265	223
◆	4%	178	149	129
··■··	6%	130	111	100

Fig. 8.3. Sensitivity analysis of the conservation fund.

Policy considerations

In reaching an appropriate amount for a conservation fund, there are various things to consider. Improvements in storage efficiencies resulting from technical change would probably lower costs in the future; but other techniques may reduce the risk of loss or genetic drift while increasing costs. Our estimates are based on data collected during a time of structural and operational change for some genebanks. Though we sought to abstract from the cost implications of these changes, on balance we are probably left with an upper-bound estimate of the relevant costs if the genebanks were to be operating in steady state.

Some factors would raise the endowment target. Our cost estimates were based on a steady-state continuation of the present level of activity into the distant future. Increasing the size of the collection or the number of samples distributed annually would obviously increase costs and the amount of funds required to support them. Conserving genetic material is a labour-intensive undertaking. If structural changes in labour markets cause local wage rates to rise faster than our assumed rate of inflation, the endowment fund would need to grow accordingly.

Moreover, our cost estimates include only those core activities required to conserve and distribute the CG holdings. Wright (1997) points out that the general lack of evaluation information on stored germplasm has severely limited its use in crop breeding and thereby curtails the demand for genebank material. Figure 8.4 provides information on the frequency with which accessions stored in the IRRI and ICRISAT genebanks were distributed during the past few decades. About 20% of the accessions held at both locations have never left the genebank. Thus, while the frequency of requests for any particular accession is low, the preponderance of the conserved material has been tapped at some point in time. Barely 2000 (2.4%) of the total of 86,720 accessions held in the IRRI collection were disseminated more than ten times during 1986–1999, while 17,800 (15%) of the total of 113,730 accessions conserved at ICRISAT were shipped out more than ten times from 1973 to 1999. In fact, 91% of IRRI's rice holdings were disseminated fewer than five times since 1986, while for ICRISAT the corresponding share is 66%, with sorghum accounting for 33% of the total shipments and pearl millet and groundnut around 19% and 13%, respectively, of the material disseminated during the 27 years since 1973.[10]

Tanksley and McCouch (1997, p. 1066) describe how modern molecular biology techniques could be used to tap the 'wide repertoire of genetic variants created and selected by nature over hundreds of millions of years [that are] contained in our germplasm banks in the form of exotic accessions'. Costing the optimal allocation of the characterization activities that provide the molecular basis for modern breeding efforts, thereby greatly enhancing conventional crop-breeding techniques, is a complex and difficult exercise, depending in part on the state and nature of the rapidly changing biotechnologies and the timing of their use (Koo and Wright, 2000).

Final Remarks

Our economic assessment of the costs of conserving plant genetic resources *ex situ* reveals the potential for reaping significant scale economies by consoli-

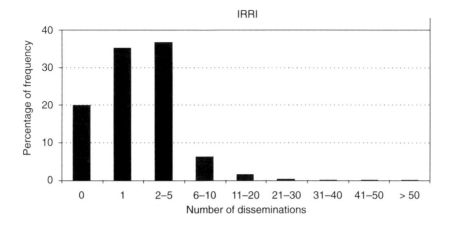

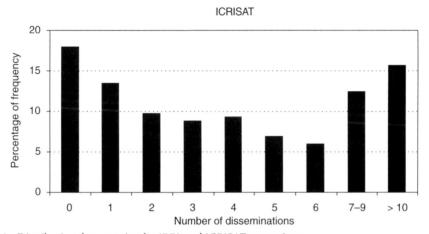

Fig. 8.4. Distribution frequencies for IRRI and ICRISAT accessions.

dating holdings within a centre (e.g. shifting from a two- to a one-module storage facility for groundnut in ICRISAT), or, perhaps, consolidating among CG centres or germplasm facilities more generally (as illustrated by the simulated costs of operating the CIMMYT genebank at current versus full capacity). While these cost savings come from spreading the use of the same or smaller amounts of physical or quasi-fixed capital over a greater number of stored accessions, conserving material over the long haul is a labour-intensive, not a capital-intensive, undertaking. Thus, lengthening the time between germination testing or regenerating material – labour-using operations that repeat at regular intervals – can significantly reduce the present value of long-run costs (e.g. adopting slower-growth *in vitro* methods for cassava at CIAT).

Notwithstanding cost savings through changes in conservation protocols or the institutional arrangements in which they are carried out, our in-perpetuity estimates show that carrying on current CG conservation practices into the far distant future is not an exorbitantly costly undertaking (especially compared with the likely benefits that can come from using this conserved material in crop-improvement programmes worldwide). However, having now assembled this core collection of crop germplasm, the funding declines of the past decade and the continuing resource uncertainties confronting CG centres (and other agencies holding key crop germplasm collections) raise real concerns over how this genetic resource is to be maintained for generations to come (Koo *et al.*, 2003b). There is an obvious and compelling case for aligning the very long-run nature of these conservation efforts with an endowment or equivalent funding mechanism, the magnitude of which we have demonstrated here.

Notes

[1]The index was formed by taking a weighted sum of the national producer price indices for the G7 countries, where the weights were the respective countries' shares of the seven-country GDP total. The indices were 100.9 in 1996, 100.6 in 1998, 101.2 in 1999 and 104.4 in 2000 (base year 1995 = 100).

[2]The costs of each operation in this chapter include fixed (capital), quasi-fixed and variable costs unless otherwise noted.

[3]The category of 'wheat' at ICARDA includes barley, whose conservation protocols and crop attributes are similar to those of wheat.

[4]This break-even cost is based solely on the savings accruing to CIAT from changing conservation protocols. Naturally, if the slow-growth method were adopted by others, the total benefits from switching to this method would increase according to spillover benefits.

[5]We considered only medium-term storage costs because groundnut accessions were not stored long-term at the time of this study.

[6]Typically, past practice seems to be that, if requests are deemed within a reasonable size, samples are distributed free of charge. But, if the amounts requested are exceptionally large, causing seed stocks to fall below threshold levels and requiring genebanks to restock their seed samples prior to shipment, then efforts are made to recover some or all of the costs involved from those to whom the seeds are shipped. This is in line with the Second Joint Statement of FAO and the CGIAR Centres on the Agreement Placing CGIAR Germplasm Collections under the Auspices of FAO.

[7]These represent the full cost of distributing seed, including the costs of maintaining the shipped samples in medium-term storage. For example, $6.63 (about 53%) of the $12.56 costs of distributing a wheat sample involves the costs of regenerating the sample, 3% for keeping the sample in medium-term storage, and only 33% covers the direct labour, handling and other variable costs involved in shipping the seed.

[8]FAO and CGIAR launched an initiative to establish the Global Crop Diversity Trust in August 2002 at the World Summit on Sustainable Development in Johannesburg. The intent was to develop an endowment fund of around $260 million, to be used to provide a permanent source of funding for the conservation of the world's important crop diversity collections held nationally, regionally and internationally. The process for the formal establishment of the Trust is expected to be completed by the end of 2004. Further details are available at <http://www.startwithaseed.org>.

[9]Specifically, we used the costs of conserving and distributing cassava at CIAT as indicators of corresponding root and tuber costs at CIP, *Musa* (banana) costs at the International Network for the Improvement of Banana and Plantain (INIBAP) and banana, cassava and yam costs at IITA. Because the methods used to conserve these

crops differ among centres, we used the *in vitro* and field genebank costs for the corresponding material held at CIP and IITA, and the *in vitro* and cryoconservation costs for the bananas stored at INIBAP. Rice costs at IRRI were deemed indicative of rice costs at WARDA; forage costs at CIAT represent forage costs at ILRI; chickpea costs at ICRISAT represent cowpea costs at IITA; and bean and forage costs at CIAT represent soybean/miscellaneous legumes and wild *Vigna* costs, respectively, at IITA.

[10]These data represent shipments from these genebanks to all sources. Nearly one-third of the material disseminated by the genetic resource unit at IRRI goes to other units within IRRI for various research purposes (Table 6.5), and around three-quarters of ICRISAT genebank material is distributed to other units within that institute. Smale and Day-Rubenstein (2002) quantified the pattern of shipments from the US National Plant Germplasm System. About three-quarters of the 621,238 accessions shipped over the past decade went to US requesters. None the less, 162,673 accessions were distributed to scientists in 191 countries and 45 territories, with nearly half these shipments going to institutions and individuals located in developing countries.

Appendix A
In-perpetuity Costs of Recurrent Expense and Annualized Cost of Capital

The in-perpetuity cost of an operation that is performed every nth year from time zero at a cost of X dollars (in other words, the present value of an item with a service life of n years purchased at time zero for X dollars and repurchased every nth year) is given by (where r is interest rate):

$$C_0^n = X + \frac{X}{(1+r)^n} + \frac{X}{(1+r)^{2n}} + \ldots = X\left[1 + \frac{1}{(1+r)^n} + \frac{1}{(1+r)^{2n}} + \ldots\right]$$

$$C_0^n = \left[\frac{1}{1-a^n}\right]X, \text{ where } a = \frac{1}{1+r} < 1. \tag{1}$$

For example, if it costs \$20 ($X = 20$) to regenerate one accession of germplasm and it is regenerated every 25 years ($n = 25$), the present value of the cost of regenerating that accession in perpetuity is \$32.00 with a 4% interest rate ($r = 0.04$).

The present value of a service costing A dollars purchased every year from time 0 is given by:

$$C_0^1 = \left[\frac{1}{1-a}\right]A, \text{ where } a = \frac{1}{1+r} < 1. \tag{2}$$

For example, if it costs \$10 to store one accession of germplasm per year, the present value of the cost of storing that accession in perpetuity is \$260 with a 4% interest rate.

To calculate the annualized user cost A of an item costing X dollars purchased every n years, we need to solve for A in terms of X by setting $C_0^1 = C_0^n$ and rearranging terms:

$$A = \left[\frac{1-a}{1-a^n}\right]X, \text{ where } a = \frac{1}{1+r} < 1. \tag{3}$$

For example, if a refrigerator costs \$2000 ($X = 2000$) and its service life is 10 years ($n = 10$), then the annualized user cost of the refrigerator is \$237 with a 4% interest rate.

The in-perpetuity cost of an operation that is performed every nth year from the kth year at a cost of X dollars is given by:

$$C_k^n = \left[\frac{a^k}{1-a^n} \right] X, \text{ where } a = \frac{1}{1+r} < 1.$$

(4)

For example, if it costs \$5 ($X = 5$) to test the viability of an accession of germplasm and it is first tested in the 10th year ($k = 10$) of introduction and then every 5 years ($n = 5$) afterwards, the present value of the cost of viability testing of one accession in perpetuity is \$18.96 with a 4% interest rate ($r = 0.04$).

Appendix B
Notes on Tables

Chapter 3: CIMMYT Genebank

All costs are in US dollars, 1996 prices. Peso-denominated costs were converted to US dollars at the rate of 7.8 pesos per dollar.

Table 3.2. Capital input costs at the CIMMYT genebank

All capital costs represent the current purchase (i.e. replacement) cost of the items involved. Some costs were not attributable to the wheat or maize operations, so they were allocated on a 50–50 basis between the two programmes. Annualized costs presented in the two right-hand columns were calculated by multiplying the replacement costs by the appropriate discount factors. The discount factors were derived using equation (3) in Appendix A, a 4% interest rate and the respective service lives. For example, the discount factors for 7, 10, 20 and 40 years of service life are 0.160, 0.118, 0.708 and 0.048 respectively.

Medium-term storage

Storage facility. Costs of constructing the medium- and long-term storage facility were taken from consolidated costing figures developed by consultants and provided by CIMMYT's finance office. Excluding the costs of shelving and refrigeration equipment, the total cost of construction is $696,204. Half of this amount was allocated to each crop, and within each crop half was allocated to medium-term and the other half to long-term storage; hence the final construction cost for medium-term storage per crop is $174,051 ($696,204 ÷ 4).

Storage equipment. Storage equipment includes the costs of refrigeration equipment and mobile shelving systems. The costs of each for one crop programme are $51,457 and $112,500, respectively, which are allocated equally between medium- and long-term storage.

Backup power system. This equipment is used mainly, but not exclusively, as a backup power option for the genebank facility in the event of a power-grid failure. The total cost of the backup power system is $41,026, but only 80% of this cost is included here (based on advice from CIMMYT facilities personnel as to the appropriate share to allocate to the genebank). This cost is allocated equally between the two crop programmes and then equally between medium- and long-term storage.

Seed container. For medium-term storage, wheat uses an aluminium bag (11 cents each); maize uses a 1 gallon plastic bucket ($2.8 each). With 123,000 accessions of wheat and 17,000 of maize, the container costs for medium-term storage are $13,530 and $47,600, respectively.

Long-term storage

Storage facility. See medium-term storage.

Storage equipment. See medium-term storage.

Backup power system. See medium-term storage.

Vacuum sealer. Each programme has one vacuum sealer ($2000) to seal aluminium packets for long-term storage (and medium-term storage for wheat).

Seed container. Wheat requires one aluminium bag (11 cents each) and maize two aluminium bags (15 cents each) for storing each accession long-term. With 75,000 accessions of wheat and 17,000 of maize, the container costs for long-term storage are $8250 and $5100, respectively.

Germination testing

Germination testing facility. The area of the germination testing room is 32 m², and the total construction cost is $12,800 ($400/m²). This room is allocated equally to each crop programme.

Germination chamber. The germination chamber is used to maintain a controlled temperature for the germination of seeds. Maize and wheat each have one chamber at a cost of $6000 each.

Other equipment. This category includes laboratory tables and chairs with a total cost of $500, which is allocated equally between the two crop programmes.

Regeneration

Screenhouse. The screenhouse is a single, stand-alone, 2000 m² structure consisting of plastic fabric on a metal frame. It is used for wheat at El Batan facility.

Vernalizer. Two vernalizers are used by the wheat programme to hold seeds at the low temperatures required to replicate cold-weather temperatures necessary for winter germination ($6000 each).

Seed-cleaning equipment. This includes a sheller for maize at a cost of $4465.

Seed-drying equipment. The wheat programme has one chamber equipped with shelves at El Batan ($25,000). The maize programme has one drying chamber for final drying at El Batan ($25,000) and two local driers for initial drying at El Batan and Tlaltizapan ($25,000 each). Given that the two regional driers are shared with the respective CIMMYT breeding programmes, the share for genebank usage is 20%.

Seed-processing facility. The area for seed processing (cleaning and packing) is 75 m² per programme at a cost of $30,000 for each.

Seed-processing equipment. This includes laboratory tables and chairs at a cost of $1500 for each programme.

Vehicles. There are two vehicles for wheat and three for maize at a cost of $13,000 per vehicle.

Seed-health testing

Seed-health testing facility. The building area occupied by the seed-health unit is 45.51 m², at a total cost of $18,204 ($400 × 45.51). Twelve per cent of this cost was attributed to genebank operations. In 1998, the seed-health unit processed a total of 51,170 accessions, including 6140 for the genebank. Sixty per cent of the cost was allocated to wheat and the rest to maize, based on the number of accessions treated (3650 of wheat and 2490 of maize).

Greenhouse. A greenhouse of 100 m² is allocated to the seed-health unit for the germination of incoming accessions. With a construction cost of $150/m², the total cost of a greenhouse is $15,000, of which 12% was allocated to the genebank on a 6:4 ratio between wheat and maize.

Laboratory/office equipment. Based on a detailed, item-by-item accounting of the laboratory equipment in the seed-health unit, we estimated that the replacement cost of this equipment totalled $145,064, of which 12% was assigned to genebank operations.

Jacuzzi equipment. The cost of this custom-built piece of equipment and its associated compressors totalled $16,794 (pool, $3717; trays, $8076; and compressor, $5000). This equipment is used to treat wheat accessions shipped from the genebank and seed samples distributed through the international wheat nursery system. Based on the average weight and distribution of seed during 1997 and part of 1999, we estimated that about 4% of the accessions treated each year are for the genebank; hence the total cost was prorated from this share.

Vehicle. Twelve per cent of one vehicle, which costs $13,000, was allocated to seed-health testing.

General capital

General facility. The area of offices and other multipurpose rooms is 62 m² for each programme, at a total of $24,800 ($400 × 62).

Office equipment. This includes furniture and fixtures such as shelving for the genebank office and workrooms at a cost of $10,000 for each programme. Computers were costed on an annual rental basis in line with CIMMYT's present charge-back procedures, and so were directly included in Table 3.3 on an annual, rental-cost-equivalent basis.

Table 3.3. Annual operating costs of conservation and distribution at the CIMMYT genebank

To estimate per-accession costs, we compiled costs for each element and its associated number of accessions for the primary survey year 1996 (as indicated below in parentheses, with the exception of seed-health costs and accession numbers, which are for 1998). Some subsequent calculations use an annual average accession figure to scale up these per-unit costs (typically an average for the 3 years 1996–1998). The base costs are net of overheads, and so the adjusted overheads rate of 22.13% was applied to each of the implicit cost-category subtotals. Representative labour costs used in our calculations are as follows:

Internationally recruited scientist: $124,600 per annum ($70,000 plus 78% fringe benefits, such as pension, health benefits and home-leave allowances)
Locally recruited specialist: $19,200 per annum ($12,000 plus 60% fringe benefits)
Secretary: $16,000 per annum ($10,000 plus 60% fringe benefits)
Assistant labour: $520 per month ($325 per month plus 60% fringe benefits)
Unskilled casual labour: $260 per month ($200 per month plus 30% fringe benefits)

Core staff for the wheat programme, all based at El Batan, includes an internationally recruited scientist (genebank head), three locally hired assistants (secondary school diplomas or equivalent) and a secretary (shared on a 50–50 basis with maize). The maize programme includes an internationally recruited scientist, a locally hired specialist and three locally hired assistants based at El Batan, plus an assistant based at Tlaltizapan. Both programmes operate prebreeding and evaluation activities, which are not included in this study. Twenty per cent of the labour for each genebank head which is spent for prebreeding and evaluation activities is excluded from this study. For wheat, we also excluded parts of the assistant labour, leaving 30 months of assistant labour included in our costing. For maize, assistants' labour for prebreeding is provided by the breeding programme, so all of the genebank assistant labour was included in our costing.

Overheads
The genebank facility is part of a more general CIMMYT operation, and thereby draws benefit from this association. Thus our cost calculations include the genebank's appropriate share of the common or 'overheads' costs incurred by CIMMYT. In 1998, CIMMYT's audited overheads rate was 30% (calculated as a loading on CIMMYT's 'operational', not total, budget). With advice from CIMMYT's director of research, we removed or reduced those cost categories included in the centre's general overheads rate that would lead to an implicit double-counting as a result of our having directly costed some of these elements. The remaining elements in the adjusted overheads rate of 22.13% used for this study are as follows:

Research support. Systems and computing services, 1.12%; biometrics, 0.76%; hardware maintenance, 0.01%; soils and plant laboratory, 0.11%; El Batan station, 1.32%.

Information services. External relations, 1.02%; donor relations office, 0.24%; publications, 1.65%; library, 1.04%.

Administration. Administration finance, purchasing, human resources, visitor services, 10.68%; plant operation, 3.03%; and depreciation, 1.15% (one-quarter of the 5.77% general depreciation allowance, representing the residual share of general depreciation charges attributable to the genebank).

Acquisition
In the survey year, 5800 accessions of wheat and 1580 of maize were acquired.

Seed-health testing. This includes costs for testing incoming accessions incurred by the Seed Health Unit (SHU). The labour cost of operating the SHU includes 50% of an internationally recruited scientist ($62,300), a locally recruited scientist ($19,200), 50% of a secretary ($8000), a technician ($16,000) and two assistants ($12,480). The labour costs of both the internationally recruited scientist and the locally recruited scientist are regarded as quasi-fixed ($81,500), while the labour cost of the remaining staff is regarded as variable ($36,480). The non-labour laboratory costs and other expenses were $42,000. In 1998, the SHU tested a total of 20,000 incoming accessions (12,750 wheat and 7250 maize) and 31,170 outgoing accessions (29,360 wheat and 1810 maize), including accessions destined for the genebank and the movement of seed in conjunction with the breeding programme. There is thought to be little, if any, difference in determining the seed-health status of a wheat or a maize accession. The testing cost of an incoming accession is about 30% higher than the cost of an outgoing accession. Based on the number of accessions tested, we allocated 45% of the labour and non-labour costs of the SHU to the testing of incoming accessions and the rest to the testing of outgoing accessions. Therefore, the quasi-fixed labour cost, variable labour cost and non-labour cost for screening an incoming accession is estimated to be $1.83 (0.45 × $81,500 ÷ 20,000 accessions), $0.82 and $0.84, respectively.

From the total number of wheat accessions introduced in the sample year (5800), about 40% were obtained from outside Mexico and were therefore subject to seed-health checks (2320 accessions). Thus for wheat the total labour costs are $6148 ([$1.83 + $0.82] × 2320 accessions) and the non-labour costs are $1949 ($0.84 × 2320 accessions). For maize the total labour costs are $4187 ($2.65 × 1580 accessions) and the non-labour costs are $1327 ($0.84 × 1580 accessions).

Introductory planting. Newly introduced wheat accessions from outside Mexico are first planted at El Batan to check for disease. A total of 2320 accessions of wheat were planted in the sample year. We used parts of the field operation costs at El Batan in Table 3.4 to calculate the field cost of introductory planting (21% = 2320 ÷ 11,000).

Seed handling. This includes activities such as registering ID numbers and sorting and packing when new samples are introduced. One month of assistant labour is required for wheat and a month of casual labour is needed for maize.

Medium-term storage

Storage management. Five per cent of the genebank head's labour is allocated to storage management, totalling $6230 ($124,600 × 0.05).

Climate control. One year of a technician's labour from the engineering unit is used to maintain the storage facility year-round, and the cost is equally allocated to medium- and long-term storage for each programme. In addition, 1 year of casual labour is assigned for cleaning and servicing the equipment. Non-labour costs represent electricity to run the equipment. The annual electricity consumption for running the medium-term storage equipment is 47,000 kWh at a cost of 6.7 cents/kWh, totalling $3149 annually. The $63 annual cost of powering the backup power system is also included here. These costs are allocated equally to each crop programme.

Long-term storage

Storage management. Three per cent of the genebank head's labour was allocated to storage management, totalling $3738 ($124,600 × 0.03).

Climate control. The labour cost for climate control under long-term storage is the same as for medium-term storage. Annual electricity consumption for the long-term storage equipment is 80,000 kWh, costing $5360. Along with the $63 annual electricity cost of running the backup power system, climate control costs are allocated equally to each programme.

Germination testing
In the survey year, 2000 accessions of wheat and 3400 of maize were tested.

Germination testing. Two per cent of each genebank head's annual labour cost was allocated to germination testing at a cost of $2492 ($124,600 × 0.02). Additional labour costs for wheat represent 4 months of assistant labour at $520 per month, plus 3 months of unskilled labour at $260 per month. The cost for maize includes 2 months of assistant labour. Non-labour costs include chemicals and other testing supplies ($200 for wheat and $100 for maize).

Dissemination
In the survey year, 14,220 accessions and 31 shipments of wheat and 3680 accessions and 69 shipments of maize were disseminated.

Dissemination management. Eighteen per cent of the genebank head's labour for managing the dissemination of accessions was allocated to wheat; 9% was allocated to maize.

Seed-health testing. Fifty-five per cent of the SHU's expenses – $44,835 in quasi-fixed labour costs, $20,064 in variable labour costs and $23,100 in non-labour costs – was divided by the number of outgoing accessions tested (31,170). The average labour cost for an outgoing accession is $1.43, $0.64 and $0.74 for quasi-fixed labour, variable labour and non-labour costs, respectively (see 'Acquisition' above). From a total of 14,220 wheat accessions distributed by the genebank, only 1330 accessions were screened by the health unit and distributed externally from CIMMYT. Thus, the seed-health costs for wheat are $2753 for labour and $984 for non-labour. There were 910 maize accessions distributed externally; hence the corresponding maize figures are $1884 and $673, respectively.

Seed treatment. The jacuzzi treatment only applies to outgoing wheat accessions. The total annual costs for treating seed in the jacuzzi were $2240 for labour and $12,253 for non-labour, of which only 4% was attributable to the genebank, based on the quantity and disposition of seed treated ($90 and $490, respectively).

Packing and shipping. One assistant can pack 500 wheat accessions per day, at a total cost of $738 ($26 × 28.4). Non-labour material costs $188 for bags (1 cent per bag × 14,220 accessions + $1.50 per box × 31 shipments), $2000 for shipping and $572 for phytosanitary certificates ($26 per certificate × 22 international shipments). One assistant can pack 150 maize accessions per day, at a total cost of $637 ($26 × 24.5). Non-labour costs include $140 for bags (1 cent per bag × 3680 accessions + $1.50 per box × 69 shipments), $5000 for shipping, and $1066 for phytosanitary certificates ($26 × 41 shipments).

Duplication
In the survey year, 35,000 accessions of wheat and 2230 of maize were duplicated.

Packing and shipping. Two per cent of the genebank head's labour for wheat and 1% for maize are allocated to packing and shipping. One assistant can process 400 wheat accessions per day at a cost of $2275 ($26 × 87.5), and 200 maize accessions per day at a cost of $299 ($26 × 11.5). Non-labour costs for wheat include the costs of storage bags (35,000 accessions at 11 cents per bag, totalling $3850) and shipping containers (36 boxes at $1.50 per box, totalling $54), plus shipping costs of $342. Maize costs include the costs of the storage bags (2230 accessions at 56 cents per bag, totalling $1249) and the shipping containers (110 boxes at $1.50 per box, totalling $165), plus shipping costs of $2000.

Information management

Database management. The labour costs of maintaining databases for wheat include 9 months of assistant labour at $4680 ($520 × 9) and 6 months of specialist labour at $9600 ($1600 × 6). The cost for maize includes 15 months of assistant labour and 6 months of specialist labour, which, at the time of this study, were all drawn from staff in the CIMMYT wheat-improvement programme.

Catalogue management. This category includes costs for additional software and database development by a CIMMYT computer programmer (2.33 months for wheat and 3.19 months for maize, charged at $1918 per month).

Other expenses. This includes supplies for data management ($500 for each crop programme).

General management

Managerial staff. This category includes labour costs for genebank staff not included elsewhere in Tables 3.3 and 3.4. Twenty per cent of the genebank head's labour is allocated here for general administration of the wheat programme and 30% is allocated for the maize programme. A secretary costing $16,000 is divided equally between the two crop programmes.

Computers. The wheat programme uses three computers, the maize programme uses four and one computer is shared by the two programmes; each computer costs $1400 per year (CIMMYT's internal charge-back rate).

Electricity. The genebank uses 50,000 kWh for offices and other laboratories, at a total cost of $2850. This cost is allocated equally between the two programmes.

Other expenses. Other expenses include office-related costs, such as the telephone, at a cost of $5000 for each programme.

Table 3.4. Annual regeneration costs at the CIMMYT genebank

Regeneration

Field operation. During the primary survey year, wheat regeneration involved a two-step process to address Karnal bunt problems. Materials were first grown at El Batan to screen the seeds for disease, but, because El Batan does not produce materials of a satisfactory quality for storage, the screened (cleaned) seed is shipped to the Mexicali area – which is free of Karnal bunt – to be planted for regeneration. In the sample year, about 11,000 accessions were grown for cleaning in a 2 ha area or in a screenhouse at El Batan; they were then regenerated in a 4 ha area at Mexicali. Maize is regenerated both at El Batan and at Tlaltizapan. Among 650 regenerated accessions in the sample year, 500 accessions were regenerated in two 2.5 ha cycles at Tlaltizapan and 150 accessions were regenerated in a 1.5 ha cycle at El Batan.

Field management. Twenty per cent of each genebank head's labour is allocated to the management of seed regeneration.

Seed preparation. One month of assistant labour is required for seed preparation for each crop. Envelopes per accession to store seed before planting cost 1 cent each.

El Batan field site. *Planting.* Wheat requires 3 months of assistant labour and 3 months of casual labour, including the labour for the screenhouse. Maize requires 1 month of assistant labour plus 2 months of casual labour. Non-labour costs for the screenhouse include fertilizer, plastic and irrigation water at $242.

Plant maintenance. Plants are cared for during the growing season by the field station unit. The estimated per-hectare cost of field management at El Batan is $1073 (apparently there is a marginal cost difference between wheat and maize, but, for the purposes of our study, we took the costs to be the same). These costs include the cost of herbicides, insecticides, fertilizers and equipment use, as well as field labour and other materials. The labour component is $333/ha. With 2 ha of wheat and 1.5 ha of maize, the labour and non-labour costs of plant maintenance are $666 and $1480 for wheat, respectively, and $500 and $1110 for maize, respectively. We also added 3 months each of assistant and casual labour for the screenhouse operation for wheat.

Pollination. To control cross-pollination, maize uses a glassine bag costing 3 cents and a pollen tector bag at 0.8 cents per plant. Each accession has 256 plants, and, with 150 accessions, there are 38,400 plants. The total cost of bags is $1459. Placing and removing bags requires 2 months of assistant labour and 4 months of casual labour.

Harvesting. Both wheat and maize require hand-harvesting. Wheat requires 4 months of assistant labour and 8 months of casual labour, and maize requires 3 months of assistant labour and 6 months of casual labour. Harvested wheat materials are put in a paper bag, costing 3 cents each, and, for maize, in a muslin cloth bag, costing 10 cents each.

Mexicali and Tlaltizapan field sites. *Planting.* One assistant flies to Mexicali for 0.2 months to supervise wheat planting. In addition, 5 months of casual labour are required for planting. Maize planting requires 2 months of assistant labour and 2 months of casual labour.

Plant maintenance. The estimated per-hectare cost of field management at Mexicali is $1217, of which $450 is labour; at Tlaltizapan the cost is $1009, of which $426 is labour. With 4 ha of wheat at Mexicali and 5 ha of maize at Tlaltizapan, the labour and non-labour costs of tending for plant are $1800 and $3068 for wheat and $2130 and $2915 for maize, respectively.

Pollination. With 500 accessions of maize regenerated at Tlaltizapan, there are 128,000 maize plants, making the total cost of bags $4864 ($0.038 × 128,000). The labour for placing and removing bags to control for cross-pollination includes 7 months of assistant labour and 15 months of casual labour.

Harvesting. Wheat requires 0.2 months of assistant labour and 30 months of casual labour; maize requires 5 months of assistant labour and 20 months of casual labour. Harvested materials are put in a paper bag at 3 cents each for wheat and in a muslin cloth bag at 10 cents each for maize.

Transportation. Transportation costs for wheat include $897 to air-freight seed from El Batan to Mexicali, plus $1410 to haul regenerated seed by truck back to El Batan, plus $1711 for the travel and related costs of El Batan staff working at the Mexicali site ($600 in plane tickets, $150 in return travel costs and $961 in per diems). The $550 cost for maize is for shipment of seed from Tlaltizapan to El Batan.

Seed processing

Processing management. Five per cent of each genebank head's labour is allocated to managing the processing of seed.

Seed cleaning/drying. The cost of labour includes 8 months of assistant labour and 12 months of casual labour, totalling $7280, for wheat. The labour cost for maize is 4 months of assistant labour plus 16 months of casual labour at Tlaltizapan, totalling $6240, and 2 months of assistant labour and 8 months of casual labour at El Batan, totalling $3120. Non-labour costs for maize include bags, envelopes and fungicides at $3500 and the electricity to run driers at $2638 for wheat and $3287 for maize.

Medium-term packing. One assistant can pack 400 accessions of wheat or 200 accessions of maize, totalling $715 ($26 × 27.5) for wheat and $85 ($26 × 3.25) for maize.

Long-term packing. The labour costs for wheat and maize are $715 and $85, respectively.

Chapter 4. ICARDA Genebank

(All costs are in US dollars, 1998 prices. Costs were converted to US dollars at the rate of 46 Syrian pounds per dollar.)

Table 4.2. Capital input costs at the ICARDA genebank

Medium-term storage

Storage facility. The area of the medium-term storage room is 122 m², so the total construction cost is $97,600 (122 × $800/m²). The refrigeration systems for both medium- and long-term storage are in a 23 m² mechanical room. Sixty per cent of this space is used for the medium-term system, so we allocated 13.8 m² to medium-term storage, at a total cost of $6900 (13.8 × $500/m²).

Storage equipment. With 48 shelves at $300 each and 2590 trays at $15 each, the total shelving and tray cost is $53,250 ($300 × 48 + $15 × 2590). Refrigeration equipment includes two dehumidifiers at $16,000 each, two cooling devices at $30,000 each and one control unit at $10,000.

Seed containers. Four different sizes of plastic containers are used for medium-term storage: very small, $0.08; small, $0.13; medium, $0.17; and large, $0.26. Cereals use medium-sized containers, with the exception of wild wheat, which uses the largest size ($10,105 = $0.17 × 52,389 + $0.26 × 4611). Food legumes use large containers, with the exception of lentil, which uses small ones ($6641 = $0.26 × 21,182 + $0.13 × 8723). Forage legumes use very small containers, with the exception of *Lathyrus*, which uses small ones ($2690 = $0.08 × 28,674 + $0.13 × 3043).

Long-term storage

Storage facility. The total construction cost for the long-term storage room, at 80 m², is $64,000; the construction cost of the mechanical room, at 9.2 m², is $4600 ($500 × 9.2 m²).

Storage equipment. The cost of shelves and trays is $37,800 ($300 × 34 shelves + $15 × 1840 trays). Refrigeration equipment includes two cooling devices and one control unit, at a total cost of $70,000 ($30,000 × 2 + $10,000).

Vacuum-sealing device. The vacuum-sealing device, at a cost of $5000, is used to seal aluminium bags for long-term storage.

Seed containers. A total of 83,197 accessions are stored in the long-term storage room (53,982 cereals, 16,727 food legumes and 12,488 forage legumes). Two to five aluminium bags at 70 cents each are used per accession, depending on the size of the seed. The cost of storage containers is $75,575 for cereals (53,982 × $0.7 × 2); $42,601 for food legumes (9135 × $0.7 × 5 + 7592 × $0.7 × 2); and $17,903 for forage legumes (12,488 × $0.7 × 2).

Viability testing

Viability-testing facility. At 46 m², the viability-testing room costs $23,000 ($500 × 46).

Incubator. The single incubator costs $12,000.

Other equipment. This category includes laboratory tables, chairs and other supplies at a total of $5000.

Regeneration

Farming equipment. Farming equipment is operated by the station operation unit. The genebank owns a scarification machine, which costs $1000 and is used to remove the hard cover from the seeds of forage legumes.

Greenhouse/screenhouse. The Genetic Resources Unit (GRU) has two greenhouses, each of 340 m². The total cost of both greenhouses is $108,800 (680 × $160 per m²). Fifteen per cent of the greenhouse area is used for food legumes, while the rest is used for forage legumes. Faba beans (food legumes) are cultivated in screenhouses to control out-pollination. The GRU owns 20 screenhouses of 4 m × 4 m each. The construction cost of one block, consisting of 24 screenhouses, was $2650 in 1989 (steel structure, $850; screen materials, $1800). Current replacement costs were calculated at $4505 for construction (adjusted at 70% inflation) and $3754 for 20 screenhouses ($4505 × [20/24]).

Seed-cleaning equipment. Cereals use three cleaning machines costing $3500, $5000 and $6000, respectively. Food and forage legumes use only one machine at a cost of $3500, for threshing and cleaning. Ten per cent of the total cost of this equipment, which is operated by the seed-production unit, was allocated to each crop.

Seed-drying facility. At an area of 11.4 m², the facility cost is $5700.

Seed-drying equipment. A dehumidifier costing $16,000 is used to dry all types of seeds.

Seed-processing facility. At a total area of 75 m², the facility costs $37,500.

Seed-processing equipment. This category includes tables, chairs and other equipment at a total cost of $2000.

Vehicles. Each group of crops – cereals, food legumes and forage legumes – uses one vehicle at a cost of $21,000 each.

Seed-health testing

Seed-health facility. The Seed Health Laboratory (SHL) building is 255 m², and the construction cost is $127,500 (255 × $500). Based on genebank operation usage, we assigned 20% of this cost to genebank-related activities.

Greenhouse. The SHL has one small greenhouse of 10 m × 8.5 m for growing out incoming materials and one large greenhouse of 40 m × 8.5 m for quarantine purposes. The construction costs are $68,000 ($160 × 425), of which 20% was allocated to the genebank.

Laboratory and office equipment. Laboratory equipment includes an incubator, vortex, centrifuge, seed treater, autoclave and related equipment, at a total capital cost of $50,000. Office equipment includes desks, chairs, photocopiers and so on, totalling $10,000. Twenty per cent of these costs were allocated to the GRU.

Computers. The SHL has three computers costing $2500 each and one printer costing $1000; 20% of these costs were assigned to the GRU.

Vehicle. One vehicle at $21,000 is assigned to the SHL, of which 20% was allocated to the GRU.

General capital

General facility. The total area of offices and multipurpose rooms is 226 m², making the total facility cost $113,000 ($500 × 226).

Office equipment. This category includes photocopiers, desks, chairs, filing cabinets, telephones and so on.

Computers. The GRU has 15 computers at $2500 each, one server at $6000, and six printers at $1000 each.

Table 4.3. Annual operating costs of conservation and distribution at the ICARDA genebank

Average staff labour costs at ICARDA are as follows:

Professional (P-level): internationally recruited staff at $92,000 per annum ($54,118 salary plus 70% employment costs including fringe benefits)
Research associate (RA-level): internationally recruited staff at $47,250 per annum ($27,794 plus 70% employment costs)
General support (GS-level): locally recruited staff at $7700 per annum ($5310 plus 45% employment costs)
Casual labour: $3.7 per day (170 Syrian pounds converted at a rate of $1 to 46 Syrian pounds)

The adjusted overheads rate used here is 24.1%, accounting for directly costed components, such as station operation, the seed unit and the physical plant unit). Currently, the (RA-level) genebank manager at ICARDA has double duty as a cereal germplasm specialist. We subtracted the cost of a specialist ($7700) from the total cost ($47,250), leaving $39,550 allocated to genebank management, of which 5% was allocated to acquisition, 20% to storage management (15% medium-term and 5% long-term), 10% to viability testing, 30% to dissemination, 5% to safety duplication and 30% to seed-processing management. The regeneration and characterization of each group of crops are managed by two crop specialists.

Acquisition

Seed-health testing. The labour cost to operate the SHL includes a part-time head (20% × $18,400), one RA ($47,250), one technician ($7700), and five casual labourers. In addition, labour to maintain the greenhouse costs $1283 (representing 2 months of technician labour). Non-labour costs include annual office and laboratory expenses of $20,000 and greenhouse operating costs of $4250 ($10 × 425 m²). Twenty per cent of all of these costs were allocated to GRU activities. Testing for incoming and outgoing accessions costs the same amount. In 1998, 2270 incoming accessions and 27,700 outgoing accessions were tested for the GRU. Hence, the labour and non-labour costs for the seed-health testing of incoming accessions are $1198 ($15,815 × [2270 ÷ 29,970]) and $367 ($4850 × [2270 ÷ 29,970]), respectively.

Seed handling. This includes 5% of the cost of the genebank manager and eight casual labourers, who can handle 300 accessions each per day.

Medium- and long-term storage

Storage management. The genebank manager allocates 20% of his time for storage management (15% for medium-term storage and 5% for long-term storage).

Climate control. The annual operating cost of the storage facility is $4504 ($12.34/day × 365 days) for labour, and $3424 for electricity ($1.34/h × 7 h/day × 365 days), which was allocated by a 3:2 ratio between the medium- and long-term storage, based on area.

Viability testing

Viability testing. It requires 1 day's casual labour to handle 50 accessions. For 6700 accessions tested, the labour cost is $496 ($3.7 × 134 days). We also added 10% of the genebank manager's labour, costing $3955 ($39,550 × 0.1). The annual non-labour cost of sponges, chemicals and Petri dishes is $200.

Dissemination

Of 132 shipments in 1998 (27,700 accessions), 75 were disseminated internally (40 to the GRU and 35 to ICARDA), three went locally to Syria and 54 were sent internationally.

Dissemination management. Thirty per cent of the genebank manager's time is dedicated to the management of seed dissemination.

Seed-health testing. The labour and non-labour costs of seed-health testing for disseminated accessions are $14,617 ($15,815 × [27,700/29,970]) and $4483 ($4850 × [27,700/29,970]), respectively. See the 'Acquisition' category above for details.

Packing/shipping. One casual labourer can pack 300 accessions per day, so 93 days of casual labour are required to pack 27,700 accessions, costing $344 ($3.7 × 93). The cost of paper bags including labels at $0.03 each is $831 ($0.03 × 27,700), and the cost of boxes for 132 shipments is $198 ($1.5 × 132). The average shipping cost is $10 for a domestic shipment and $40 for an international shipment totalling $2190 (3 × $10 + 54 × $40). There were 54 international shipments requiring phytosanitary certificates at $108 ($2 each × 54).

Duplication

Packing/shipping. Five per cent of the genebank manager's labour is allocated to the management of packing and shipping, costing $1978 ($39,550 × 0.05). A casual labourer can handle 300 accessions per day; hence the total labour cost to pack 8410 accessions is $103 ($3.7 × 28 days). Non-labour costs include aluminium bags at 70 cents each, totalling $5887 ($0.7 × 8410), labels costing $84 ($0.01 × 8410), and boxes costing $59 (39 × $1.5). In 1998, there were five shipments: lentil to India ($203), chickpea to India ($847), *Lathyrus* to Switzerland ($85), *Medicago* to Austria ($77) and *Vicia* to Austria ($199), representing a total shipping cost of $1411.

Information management

Database management. Labour information management includes one RA and one assistant. Non-labour costs include a software licence fee at $3000. Since these costs include prebreeding and evaluation activities, only 80% of these costs were allocated to the genebank costing.

Other expenses. The cost of publication is $4000, of which 80% was allocated here.

General management

Managerial staff. This category includes the genebank head, one assistant, and one casual labourer for 1 year ($888). Sixty per cent of the genebank head's labour and 80% of the assistant and casual labour costs were allocated here.

Electricity. Eighty per cent of the annual electricity cost – other than electricity for storage – is allocated under this category, for a total of $3424.

Other expenses. This includes the telephone ($600), e-mail ($2400), office and computer supplies ($3500) and other miscellaneous costs ($2000), of which 80% is included here.

Tables 4.4 and 4.5. Annual costs of regeneration and characterization at the ICARDA genebank

Regeneration

Field management. Two (GS-level) germplasm specialists oversee the regeneration and characterization of each crop. Eighty per cent of the labour was allocated to field management for regeneration, and 20% to recording traits for characterization. Annual diesel fuel consumption of $500 is allocated equally to each crop.

Seed preparation. Fungicide is applied to seeds before planting. Cereals require 20 days of casual labour, costing $74 (3.7 × 20); forage legumes require 30 days; chickpea, 5 days; lentil, 5 days; and faba bean, 10 days. The fungicide costs are $50 for cereals, $70 for forage legumes, $10 for chickpea, $10 for lentil and $25 for faba bean. In addition, wild lentil (400 accessions) needs hand-scarification, costing $14.8 ($3.7 × 4 days of casual labour), and half of the forage legumes (3500 accessions) need machine-scarification costing $130 ($3.7 × 35 days of casual labour).

Land preparation. The station operation charges $400/ha for land preparation. The costs are $1600 for cereals (4 ha), $1400 for forage legumes (3.5 ha), $400 for chickpea (1 ha), $400 for lentil (1 ha), and $1000 for faba bean (2.5 ha). Seventy per cent of these costs were allocated to labour and 30% to non-labour.

Planting. Cereal requires machine planting, using 10 days of casual labour at $37 ($3.7 × 10); forage legumes require 90 days of casual labour both in the field and in greenhouses, costing $333 ($3.7 × 90). For food legumes, lentils are planted by a machine called 'Ojord' requiring 2 days of casual labour and one driver at a cost of $17 ($10 × 1 + $3.7 × 2); chickpeas are planted by a machine called 'Wintersteiger' requiring 2 days of casual labour and one driver at a cost of $17 ($10 × 1 + $3.7 × 2); and faba beans are hand-planted, requiring 20 days of casual labour at a cost of $74 ($3.7 × 20).

Screenhouse. The labour and non-labour costs of operating the greenhouse, at 680 m², are $1282 and $680, respectively. Fifteen per cent was allocated to lentil and 85% to forage legumes. The labour and non-labour costs of operating the screenhouse, at 320 m², are $641 (representing 1 month of a technician's labour) and $320, respectively. The screenhouse is exclusively used for faba bean.

Chemicals and hand-weeding. Labour for spraying chemicals is included in the cost of land preparation. The chemical cost of fertilizers and herbicides is $142 for cereals, $2372 for forage legumes and $1816 for food legumes, all allocated by land area. For hand-weeding, cereals require 80 days of casual labour, costing $296; forage legumes require 100 days, costing $370; and chickpea and lentil require 10 days each, costing a total of $74.

Harvesting. Cereals require 12 days of casual labour and one driver, costing $84; forage legumes require hand-harvesting, using 160 days of casual labour (for both the field and the greenhouses), costing $592. Chickpea is machine-harvested, requiring 10 days of casual labour and one driver, costing $87. Both lentil and faba bean are hand-harvested, requiring 15 and 30 days of casual labour, costing $56 and $112, respectively. Harvested materials are transported either in paper bags at 2 cents each for cereals or in cotton bags at 22 cents each for legumes.

Seed processing

Processing management. Thirty per cent of the genebank manager's time was allocated to the seed-processing management. This cost was allocated to each crop based on the number of accessions processed.

Seed cleaning. Cereals require both hand- and machine-cleaning. Machine-cleaning requires 60 days of casual labour, costing $222; hand-cleaning requires 45 days of casual labour, costing $167. Forage legumes require 232 days of casual labour, costing $858. Both chickpea and lentil are machine-cleaned, requiring 10 days of casual labour each, at a cost of $74. Faba bean is hand-cleaned, requiring 80 days of casual labour at a cost of $296.

Seed drying. Weighing and separating seed prior to drying is done by casual labourers, costing $53 for cereals, $81 for forage legumes, $11 each for chickpea and lentil and $20 for faba bean.

Medium-term packing. One casual labourer can process 250 accessions per day.

Long-term packing. One casual labourer can process 200 accessions per day.

Characterization

Recording traits. Germplasm specialists record traits in the field, representing 20% of their time, assisted by casual labour. Cereals require 100 days of casual labour; forage legumes, 200 days; chickpea and lentil, 30 days each; and faba bean, 60 days.

Chapter 5: ICRISAT Genebank

(All costs are in US dollars, 1999 prices. Costs were converted to US dollars at the rate of 45 rupees per dollar.)

Table 5.2. Capital input costs at the ICRISAT genebank

Medium-term storage

Storage facility. The building area of the medium-term storage facility is 470 m^2, costing an average of $250/m^2 for construction for a total of $117,500.

Storage equipment. Six storage modules (four medium and two large) are installed inside the building, each equipped with a cooling system, a compressor and two dehumidifiers. The cost of a medium-sized module, at 9.1 m × 5.1 m, was 18,000 GBP in 1983. This was converted to a 1999 purchase price in US dollars using the UK manufacturing output price index and the current exchange rate published by the International Monetary Fund. The resulting cost of each module is $50,234. The cost data of a large-sized module, at 10.1 m × 7.6 m, were unobtainable, so we calculated the purchase cost using the data for the medium-sized module. Given that the large-sized module is 1.65 times bigger than the medium-sized one, the current purchase price is estimated to be $83,084 (1.65 × $50,234). Both mobile and fixed shelving systems are installed in the medium-term storage modules, amounting to $21,125 for mobile and $12,316 for fixed systems. The total tray cost is $18,100, representing 3000 aluminium trays at $5.30 and 1000 plastic trays at $2.20.

Other equipment. A hygrothermograph costing $390 is installed in each module to check the relative humidity. A safety system, costing $632, is centrally located to check all the modules; and three automatic temperature control systems, costing $391 each, are available as backup. Two-thirds of these costs were allocated to medium-term storage ($1203 = [$632 + $1173] × [2/3]) and one-third to long-term storage ($602 = [$632 + $1173] × [1/3]).

Seed container. Three types of aluminium cans and a plastic jar are used for medium-term storage. Pigeonpea uses large (7.5 cm × 15 cm) aluminium cans, costing $1.18 each, small millet uses small (5 cm × 7 cm) aluminium cans, costing $0.55 each, groundnut uses large (1500 ml) plastic jars, costing $0.25 each, and the remaining accessions use medium (7.5 cm × 10 cm) aluminium cans, costing $0.96 each. Thus, the container cost for each crop is $35,232 for sorghum ($0.96 × 36,700), $20,544 for pearl millet ($0.96 × 21,400), $16,608 for chickpea ($0.96 × 17,300), $15,930 for pigeonpea ($1.18 × 13,500), $3850 for groundnut ($0.25 × 15,400) and $5060 for small millet ($0.55 × 9200), amounting to a total of $97,224. In addition to the containers, a metal name-plate on the surface of each can labels each accession. The name-plates were purchased in bulk in 1984 at a cost of $11,944; the current replacement cost is calculated to be $14,619.

Long-term storage

Storage facility. The total construction cost of long-term storage, at 216 m^2, is $54,000 ($250 × 216).

Storage equipment. Three modules at $50,234 each are installed in the building. The costs of mobile and fixed shelving systems are $21,830 and $2145, respectively; the trays cost $2200 (1000 × $2.2).

Other equipment. This category includes a third of the cost of the safety and automatic systems ($1805 × 1/3), a sign-maker for labelling trays ($347) and two vacuum-sealing machines ($3080 each).

Seed container. Three different sizes of aluminium packets are used for long-term storage, at an average cost of 15 cents. One bag is used for each crop, with the exception of groundnut, which uses two bags. The total container cost for long-term storage is $19,335.

Viability testing

Viability-testing facility. The viability-testing facility, at 96 m^2, costs $24,000.

Seed germinator. Two seed germinators, costing $9890 each, are used for germination testing.

Other equipment. This category includes two incubators, costing $6570 each, a water distiller, at $3420, and other furniture and equipment, at $5000.

Regeneration

Greenhouse. Wild groundnut is regenerated in a greenhouse that is divided into three bays of 117 m^2 each. Two bays are used for regeneration; the remaining bay is used as a field genebank for wild groundnut. With an average construction cost of $100/m^2, the cost of the facility is $23,400 for regeneration (two bays). There are six coolers at $950 each and 18 mobile benches in each bay at $50 each. Each bench holds 16 pots costing $2.20 each that last for 5 years. Thus, the total cost of benches is $1800 ($50 $\times$ 18 $\times$ 2 bays) and the cost of pots is $25,344 ($4.4 $\times$ 288 $\times$ 2 bays), assuming the pots need to be replaced once over the greenhouse's service life.

Seed-cleaning equipment. This includes a seed blower at $2240, a seed counter at $4630, a seed divider at $2030 and a precision scale at $974.

Seed-drying facility. The building area for the seed drying cabinet is 72 m^2 and costs $18,000 (72 $\times$ $250).

Seed-drying equipment. The seed-drying cabinet, equipped with cooling system, dehumidifiers and shelving system, costs $60,040.

Seed-processing facility. The area of the seed-processing room (short-term storage room) is 190 m^2, amounting to a construction cost of $47,500 (190 $\times$ $250).

Seed-processing equipment. The seed-processing room is equipped with eight small dehumidifiers (five operational and three as backup), two air-conditioners at $1000 each, an exhaust system at $3160, and other furniture costing $2000.

Seed-health testing
Of 15,000 accessions tested for seed health by the Plant Quarantine Unit (PQU) in 1999, 12,000 accessions were for dissemination, of which 3195 were from the genebank. The 3000 accessions not for dissemination were for long-term storage in the genebank. Testing of disseminated materials for the breeding programme cost about twice as much as the materials for genebank and is similar to the cost for long-term storage. Based on these calculations, we allocated 33% of the PQU activity to the genebank, one-third to dissemination and two-thirds to long-term storage.

Seed-health testing facility. The service area of the building is 204 m^2 (including laboratories and offices), so the total building cost is $51,000 (204

× $250), and the construction cost of the fumigation facility, at 20 m², is $3600 (20 × $180). Thirty-three per cent of these costs were allocated to the genebank.

Greenhouse. The construction cost of the greenhouse, at 8 m × 7 m, is $5600 (56 × $100). There are three coolers in the greenhouse costing $950 each. Thirty-three per cent of these costs are allocated to the genebank.

Laboratory/office equipment. Laboratory equipment includes microscopes, an air purifier, an X-ray machine and a seed treater. The total cost of this equipment is $100,000. The office equipment includes desks, chairs and filing cabinets, costing a total of $10,000. Thirty-three per cent of this cost is allocated to the genebank.

Computers. Equipment includes four PCs with uninterrupted power system (UPS) devices at $1620 each and two printers at $700 each, of which 33% is allocated to the genebank.

General capital

General facility. The area of the remaining offices and other multipurpose rooms in the genebank is 348 m², costing $87,000 ($250 × 348).

Office equipment. This includes desks, tables, chairs, telephones, filing cabinets, copier and so on, at $20,000.

Computers. There are two servers with UPS devices costing $7900 each; 12 PCs with UPS devices costing $1620 each; eight printers costing $700 each; and other computer-related equipment costing $1000.

Table 5.3. Annual operating costs of conservation and distribution at the ICRISAT genebank

The average annual labour costs in 1999, including salary and benefits, are as follows:

> Internationally recruited scientist (IRS), $86,800
> Senior management group (SMG), $14,800
> Scientific officer (SO), $7000
> Research technician (RT), $3600
> Regular workforce (RWF), $2000

The daily wage of a temporary farm labourer (TFL) is 52 rupees, or $1.20 (based on an exchange rate of 45 rupees to US$1). The genebank unit is staffed by one IRS (the genebank head), three SMG (a genebank manager and two senior germplasm scientists), five SOs (one for genebank operation and four for regeneration and characterization), 24 RTs and 14 RWF. The staff above SO level are classified as quasi-fixed capital (or human capital) costs. Currently, the genebank head also works as a senior germplasm scientist, so we subtracted the cost of an SMG from the cost of the genebank head. Apart from research, 40% of the genebank head's time is spent on conservation-related genebank management ($28,800), as is 80% of the genebank manager's time ($11,840) and 20% of each of the senior germplasm scientists' time for each crop ($2960 each). The assumed overheads rate is 22.4%.

Acquisition

Though there were no new acquisitions in the sample year, we used average figures for a typical year. No seed-health testing is required because that service is provided by the Indian government.

Seed handling. Two per cent of the genebank manager's time and all of a scientific officer's time are spent for the management of acquisition, costing $377 ([$11,840 + $7000] × 0.02). A technician can handle about 50 incoming accessions per day. A storage bag costs 1 cent per accession.

Medium-term storage

Storage management. Fifteen per cent of the genebank manager's time and all of a scientific officer's time are spent on the management of the medium-term storage.

Climate control. On average, one technician (from the farm and engineering service programme and the genebank) is required to monitor and operate the storage facility year-round, along with $2000 in material costs. Seventy-five per cent of these costs were allocated to medium-term storage and 25% to long-term storage, based on area. Electricity to run one large-sized and one medium-sized storage module are 120,000 and 80,000 units, respectively. With two large-sized and four medium-sized modules for medium-term storage, annual electricity consumption is 560,000 units at $0.12 per unit, totalling $67,200.

Long-term storage

Storage management. Five per cent of the genebank manager's time and all of a scientific officer's time are spent for the management of the long-term storage.

Climate control. Twenty-five per cent of a technician's time, along with annual material costs, was allocated to long-term storage. Only two of the three modules were operated in 1999, so the total electricity cost for long-term storage was $19,200 ($0.12 × 80,000 × 2).

Viability testing

Viability testing. Fifteen per cent of the genebank manager's time and all of a scientific officer's time were allocated to the management of viability testing, given active testing in the sample year. Two technicians and 50% of an RWF conducted viability testing year-round, costing $8200 ($3600 × 2 + $2000 × 0.5). A paper towel costing 6.9 cents is used in testing each accession, amounting to $2070 for 30,000 accessions. In addition, the annual cost of supplies and chemicals is $200.

Dissemination

Dissemination management. Twenty per cent of the genebank manager's time and all of a scientific officer's time are allocated to dissemination. Germplasm scientists help locate requested materials, so 10% of their time is also allocated to this activity.

Seed-health testing. The labour in the plant quarantine unit includes 50% of a senior manager ($7400), three scientific officers ($21,000), five technicians ($18,000) and four RWF ($8000), making a total quasi-fixed labour cost of $28,400 and a labour cost of $26,000. Annual operating costs of the PQU are $15,000. Based on the number of accessions tested, we allocated 11% of these costs to dissemination.

Packing/shipping. A half-time technician oversees the dissemination and duplication of all materials, with the ratio between them based on the number of accessions processed. One hundred accessions can be packed per day by one RWF; hence the total labour cost to pack 21,400 accessions is $1712 (daily wage of $8 × 214). Aluminium packets for disseminating accessions cost 15 cents each, amounting to $3210 in total ($0.15 × 21,400). No cost is incurred in obtaining phytosanitary certificates. With 163 shipments at an average cost of $10, the total shipping cost is $1630.

Duplication

Packing/shipping. Three per cent of the genebank manager's time and all of a scientific officer's time were allocated here. The cost of labour to pack accessions is $174, representing a half-time technician and additional temporary labour (0.5 × $3600 × (1080/22,480) + $8 × 11). The total cost of aluminium bags is $162 ($0.15 × 1080). The estimated shipping cost is $500.

Information management

Database management. Ten per cent of the genebank manager's time and all of a scientific officer's time were allocated to information management, along with 1.5 technicians to manage the data. The licensing fee to use the SQL server is $4950.

Other expenses. This includes miscellaneous equipment such as magnetic tapes and jazz and zip disks, totalling $1000.

General management

Managerial staff. Forty per cent of the genebank head's time is spent on general management, costing $28,800 ($72,000 × 0.4), along with one secretary and half a technician's time, totalling $5400 ($3,600 × 1.5).

Electricity. In 1999, 924,000 units of electricity were consumed by the genebank. Excluding the consumption by the storage facility and seed-drying room, consumption was 124,000 units, totalling $14,880 ($0.12 × 124,000).

Other expenses. The cost of miscellaneous supplies is about $1000. One vehicle is allocated for the genebank's general use at a monthly charge of $105 plus $80 in fuel.

Tables 5.4 and 5.5. Annual costs of regeneration and characterization at the ICRISAT genebank

Regeneration

Field operation. Regeneration is broadly classified as field operation and seed processing. Three senior germplasm scientists manage the regeneration and characterization of two crops each, and four scientists undertake cultivation in the field. Excluding research, we allocated 20% of a senior scientist's time for the management of each crop, costing $2960 ($14,800 × 0.2). For groundnut, 18% of this labour was allocated to cultivated groundnut and the remaining 2% to wild groundnut. Ninety per cent of each scientist's time was allocated to field management, costing $6300 ($7000 × 0.9). The time allocated to regeneration and characterization for each crop is as follows:

> Sorghum: 1 scientific officer, 3 technicians and 1 RWF
> Pearl millet: 1 scientific officer, 2 technicians and 2 RWF
> Pigeonpea: 0.4 scientific officers, 1 technician and 1 RWF
> Chickpea: 0.5 scientific officers, 2 technicians and 2 RWF
> Groundnut: 0.5 scientific officers, 2 technicians and 2 RWF
> Wild groundnut: 0.25 scientific officers, 1.5 technicians and 0.7 RWF

Sorghum is characterized both in the rainy season and in the post-rainy season (in the same field as the regeneration). Fifty per cent of a scientist's time was allocated to field management for regeneration, 40% to characterization in the post-rainy season and 10% to field management and characterization in the rainy season.

Pearl millet is characterized both in the rainy season and in the post-rainy season (in a separate field). Forty per cent of a scientist's time was allocated to field management to regenerate 660 accessions. For 1250 accessions, 20% of a scientist's time was allocated to the characterization of four traits in the post-rainy season (10% for field management and 10% for recording traits) and 40% of a scientist's time was allocated to characterization of 20 traits in the rainy season (10% for field management and 30% for recording traits).

Pigeonpea is characterized in the post-rainy season only, in the same field as regeneration. Twenty per cent of a scientist's time was allocated to field management and another 20% to recording traits for 600 accessions of pigeonpea.

Chickpea is characterized in the post-rainy season only, in a separate field from regeneration. Thirty per cent of a scientist's time was allocated to field management to regenerate 2400 accessions; 20% of a scientist's time was allocated to characterizing 2000 accessions (10% for field management and 10% for recording traits).

Groundnut is characterized both in the rainy season and in the post-rainy season in a separate field. Twenty per cent of a scientist's time was allocated to field management to regenerate 1300 accessions, 20% was allocated to the field management and characterization of 1500 accessions in the rainy season, and 10% was allocated to the field management and characterization of 600 accessions in the post-rainy season.

Wild groundnut is characterized in the post-rainy season. For 100 accessions, 20% of a scientist's time was allocated to regeneration and 5% was allocated to characterization.

Field management. Each crop requires 20% of a senior germplasm scientist's time and parts of a scientist's time (as described in 'Field operation' above).

Seed preparation. Seeds need to be identified, retrieved, sorted, put into envelopes and clearly labelled before planting. The labour required to perform these functions for each crop is as follows:

> Sorghum: 0.15 technicians and 0.05 RWF
> Pearl millet: 0.05 technicians or RWF
> Pigeonpea: 0.05 technicians and 0.05 RWF
> Chickpea: 0.05 technicians, 0.05 RWF and 200 TFLs
> Groundnut: 0.05 technicians, 0.05 RWF and 400 TFLs
> Wild groundnut: 0.05 technicians and 0.05 RWF

Seed preparation for wild groundnut also requires scarification to remove the hard outer shell from seed and germination in a germination chamber. An envelope for each accession costs 1 cent on average.

Land preparation. The field is maintained by the Farm and Engineering Service Program (FESP), which charges field-maintenance fees. This charge includes both the labour and non-labour costs of tillage, irrigation, chemicals, farm maintenance and weeding. Excluding chemicals and weeding/bird-scaring labour, which is directly costed, the net cost of land preparation (including fertilizer) per hectare is $555 for sorghum, $609 for pearl millet, $813 for pigeonpea, $602 for chickpea, $790 for groundnut and $660 for the management of the greenhouse for wild groundnut. We allocated 70% of this charge to labour and the balance to non-labour. In addition, genebank staff conduct field layout and other land-preparation activities prior to planting. For example, both chickpea and pigeonpea require solarization to control soil-borne diseases during summer, and wild groundnut requires soil mixture (costing $88) and pasteurization. The labour cost for this operation is as follows:

> Sorghum: 0.15 technicians and 0.05 RWF
> Pearl millet 0.05 technicians and 0.05 RWF
> Pigeonpea: 0.05 technicians and 0.05 RWF
> Chickpea: 0.1 technicians and 0.1 RWF
> Groundnut: 0.05 technicians and 0.05 RWF
> Wild groundnut: 0.05 technicians and 0.05 RWF

Planting. Planting can be done either mechanically using a tractor-drawn planter or manually, depending on the crop. The tractor charges and driver's time are included in the land-preparation cost charged by the FESP. The labour required from the genebank is as follows:

> Sorghum: 0.15 technicians and 0.05 RWF
> Pearl millet: 0.05 technicians and 0.05 RWF
> Pigeonpea: 0.08 technicians, 0.08 RWF and 40 TFLs
> Chickpea: 0.15 technicians, 0.15 RWF and 200 TFLs
> Groundnut: 0.1 technicians, 0.1 RWF and 200 TFLs
> Wild groundnut: 0.05 technicians and 0.05 RWF

Plant maintenance. This includes thinning/filling, weeding/bird scaring, field labelling and roguing/purification. The labour required for thinning and filling is as follows:

> Sorghum: 0.25 technicians and 0.05 RWF
> Pearl millet: 0.05 technicians, 0.05 RWF and 45 TFLs
> Pigeonpea: 0.02 technicians, 0.02 RWF and 10 TFLs

Weeding and bird scaring are done by temporary workers employed by the FESP under the supervision of genebank staff. The labour cost for the genebank and the FESP is as follows:

> Sorghum: 0.05 technicians, 0.1 RWF and $30 × 4 ha
> Pearl millet: 0.05 technicians, 0.05 RWF, 140 TFLs and $54 × 1 ha
> Pigeonpea: 0.03 technicians, 0.03 RWF and $139 × 0.5 ha
> Chickpea: 0.05 technicians, 0.05 RWF and $37 × 1.5 ha
> Groundnut: 0.05 technicians, 0.05 RWF and $162 × 1 ha

We also included the cost of chemicals in this category under non-labour as follows:

> Sorghum: $99 × 4 ha
> Pigeonpea: $95 × 0.5 ha
> Chickpea: $64 × 1.5 ha
> Groundnut: $113 × 1 ha

The labour cost for field labelling is as follows:

> Sorghum: 0.1 technicians and 0.1 RWF
> Pearl millet: 0.05 technicians and 0.05 RWF

Pigeonpea: 0.02 technicians and 0.02 RWF
Chickpea: 0.05 technicians and 0.05 RWF
Groundnut: 0.05 technicians and 0.05 RWF

Materials used for labelling include bamboo pegs, at 3.3 cents each, and two labels, at 1.7 cents each (3.4 cents per peg). On average, one bamboo peg is used for five accessions. Since bamboo pegs last 2–3 years, the cost of a peg and label per five accessions is estimated to be 5 cents, or 1 cent per accession. The labour cost for roguing and purification is as follows:

Sorghum: 0.15 technicians and 0.1 RWF
Pearl millet: 0.1 technicians/RWF
Chickpea: 0.3 technicians and 0.3 RWF
Groundnut: 0.2 technicians and 0.3 RWF
Wild groundnut: 0.25 technicians and 0.25 RWF

The annual water cost for the greenhouse is $176 ($88 × 2 bays) and the electricity cost of running coolers is $5040 ($2520 × 2 bays).

Selfing. To prevent cross-pollination, some crops need to be covered with bags. The labour cost for selfing is as follows:

Sorghum: 1 technician, 0.25 RWF and 2960 TFLs
Pearl millet: 0.3 technicians, 0.3 RWF, and 170 TFLs
Pigeonpea: 0.25 technicians, 0.25 RWF and 300 TFLs

Sorghum requires 50 paper bags costing 0.78 cents per accession, totalling $1950 ($0.39 × 5000); the cost of stapling is $46. Pearl millet uses 160 paper bags at 4 cents per accession, totalling $4224 ($6.4 × 660); the cost of stapling is $43. Each accession of pigeonpea uses 45 muslin bags at 56 cents each, which lasts two seasons, making the cost per accession $12.6 ($0.28 × 45) and the total cost $7560 ($12.6 × 600).

Harvesting/threshing. The labour requirement for harvesting is as follows:

Sorghum: 0.3 technicians, 0.1 RWF and 300 TFLs
Pearl millet: 0.1 technicians, 0.1 RWF and 75 TFLs
Pigeonpea: 0.2 technicians, 0.2 RWF and 150 TFLs
Chickpea: 0.2 technicians, 0.2 RWF and 600 TFLs
Groundnut: 0.15 technicians, 0.15 RWF and 600 TFLs
Wild groundnut: 0.1 technicians and 0.1 RWF

Harvested materials are put into a gunny bag, costing 16 cents each, with labels at 3.4 cents per bag and ties at 2 cents each, amounting to 21.4 cents per bag. After harvesting, materials are threshed and cleaned, requiring the following labour components:

Sorghum: 0.3 technicians, 0.1 RWF and 240 TFLs
Pearl millet: 0.1 technicians, 0.1 RWF and 125 TFLs
Pigeonpea: 0.1 technicians, 0.1 RWF and 100 TFLs
Chickpea: 0.1 technicians, 0.1 RWF and 400 TFLs
Groundnut: 0.1 technicians, 0.1 RWF and 400 TFLs
Wild groundnut: 0.1 technicians and 0.1 RWF

Threshed materials are put in a muslin bag costing 4 cents each. Miscellaneous costs include one vehicle ($105 per month × 3 months), two motor bikes (2 × $6 per month × 3 months), fuel costs ($80 per month × 3 months), and other miscellaneous items ($100). Three months of a driver's time, at $900, is also allocated to each crop. The vehicle cost was apportioned to field operations for characterization, where applicable.

Seed processing

The genebank head and a scientist manage all the seed-processing operations including drying, cleaning and packing for all crops, in addition to other operations considered in Table 5.3. Thirty per cent of their time is allocated to each crop, based on the number of accessions processed.

Seed cleaning. Fifty per cent of a technician's time to supervise seed-cleaning processes for all crops is allocated based on the number of accessions regenerated. One temporary worker can shell and clean 15 accessions of groundnut per day and process 20 accessions for other crops.

Medium-term packing. One technician can pack 50 groundnut accessions or 100 accessions for other crops per day.

Seed drying. Only accessions for long-term storage are dried in the seed-drying room. Thirty per cent of a technician's time is needed to dry seed for all crops. Electricity to operate the seed-drying room amounted to 80,000 units at a cost of $9600 ($0.12 × 80,000). These costs were allocated based on the number of accessions processed.

Long-term seed-health testing. Twenty-two per cent of the operating cost of the PQU is allocated to seed-health testing for long-term storage. In 1999, 3000 accessions were tested for seed health, amounting to $6248 in quasi-fixed costs, $5720 in labour and $3300 in non-labour costs. Consequently, the per-accession cost of seed-health testing is $2.08 in quasi-fixed costs, $1.91 in labour and $1.10 in non-labour costs. This was multiplied by the number of accessions regenerated.

Long-term packing. One technician can pack 10 groundnut accessions or 20 accessions of other crops per day. The total packing cost was estimated based on these figures.

Characterization

While sorghum, pigeonpea and wild groundnut use the regeneration field for characterization in the post-rainy season; pearl millet, chickpea and groundnut require a separate field operation for characterization in the post-rainy season. Sorghum, pearl millet and groundnut need additional characterization in the rainy season.

Chapter 6: IRRI Genebank

(All costs are in US dollars, 1999 prices. Costs were converted to US dollars at the rate of 44 pesos per dollar.)

Table 6.2. Capital input costs at the IRRI genebank

Medium-term storage

Storage facility. The construction cost of the medium-term storage room is $124,124 based on 182 m^2 at $682/m^2 (30,000 pesos/m^2). The cooling systems for both medium- and long-term storage are housed in a separate area of 258 m^2 (12 m × 21.5 m). The construction cost of this area, at $170/m^2, is $43,860. We allocated 60% of the construction cost to the medium-term storage based on the space occupied.

Storage equipment. There are three cooling systems, each with a condensing unit and evaporation unit, costing $12,240. One display monitor costs $3220, and four dehumidifiers cost $3760 each. In addition, there are 34 movable shelves each costing $2500, and 3400 steel trays at $8 each.

Heat-sealing device. There are two heat sealers at $1220 each, used to seal aluminium bags.

Seed container. One large (11 cm × 15 cm) aluminium bag, costing 29 cents, is used for medium-term storage. With 86,800 accessions, the total container cost is $25,172.

Long-term storage

Storage facility. The total construction cost of the long-term storage room, at 57 m^2, is $38,874 ($682 × 57). Forty per cent of the building, used for the cooling system, was also allocated here at a cost of $17,544 ($43,860 × 0.4).

Storage equipment. This includes two cooling systems at $14,670 each, one display monitor at $3220, 16 movable shelves at $2500 each and 2300 trays at $8 each.

Vacuum-canning device. This includes a vacuum pump at $2050 and a canning machine at $12,300.

Seed container. Two aluminium cans, costing 20 cents each and containing about 50 g of rice, are used for each accession in long-term storage. With 83,930 accessions, the total container cost is $33,572.

Viability testing

Viability-testing facility. The area of the seed-testing room is 72 m^2 representing a total construction cost of $24,480 ($340 × 72).

Germination chamber. Five chambers costing $21,200 each are used for viability testing.

Other equipment. This includes two ovens at $3190 each to break seed dormancy, one oven at $2160 for moisture testing, two seed grinders at $60 each, five desiccators at $490 each, two scales at $2460 each and a moisture tester at $1000, for a total of $17,030.

Regeneration

Farming facility. A separate building of 500 m^2 to hold farm equipment in the field cost $85,000 to construct ($170 × 500). Ninety per cent of this cost was allocated to the genebank operation. (The remaining 10% is allocated to INGER activities.)

Farm equipment. This includes an air-blast cleaner at $1700, a hydro-tiller at $490, a moisture tester at $200, a power weeder at $400, four seed separators at $1800 each, two threshers at $4000 each, 12 seed blowers at $2600 each, and other equipment, such as ladders and tables, at $500. Again, 90% of these costs were allocated to the genebank.

Screenhouse. A screenhouse of 4000 m^2 is used to regenerate wild and cultivated rice, which needs special care. The construction cost including screens is $50/m^2, amounting to $200,000 in total ($50 × 4000). Severe weather conditions in the area require the screen to be replaced every 5 years at a cost of $22,400. The cost of one replacement was added to the cost of the screenhouse at 4% interest ($200,000 + $22,400 × [1/(1 + 0.04)^4]). Sixty per cent of the area is dedicated to wild rice and 40% to cultivated rice. Three pots at $1.93 each are used to multiply one accession of wild rice; with 500 accessions regenerated in the screenhouse, the total cost of pots is $2895.

Embryo-rescue facility. The area of the Conservation Support Laboratory in which *in vitro* embryo rescue is conducted is 60 m^2, with the total cost at $20,400. Since 30% of activity in the lab is related to the conservation-related activity for regeneration, we allocated 30% of the facility cost here. In 1999, a total of 1300 accessions went through this operation, of which 500 accessions were wild rice. So we allocated 40% of the cost to wild rice and the rest to cultivated rice.

Embryo-rescue equipment. This includes laboratory equipment (such as scales, dispensers, oven, sterilizers), furniture (such as laboratory tables, chairs, shelves) and durable supplies (such as scissors, spatulas and trays); the total cost is $49,330, of which 30% was allocated to the genebank.

Seed-processing facility. The area of the seed-processing room is 192 m², at a total cost of $65,280.

Seed-cleaning equipment. Inside the seed-processing room, there are four seed blowers costing $2100 each, a dust collector at $5950, and other miscellaneous equipment including tables and chairs at $2000.

Seed-drying facility. The area of the seed-drying room is 56 m², costing $19,040 (56 × $340); the machine room that houses equipment is 10 m² and costs $1700 (10 × $170).

Seed-drying equipment. The cost of seed-drying equipment was $116,669 in 1993, including a compressor, a dehumidifier, shelves and other necessary systems. Using the US Industrial Goods Producer Price Index, the current purchase price is estimated at $125,536.

Vehicles. Two pick-up trucks at $12,000 each are assigned to multiplication activities.

Seed-health testing
In 1999, the SHU tested 12,195 incoming and 57,720 outgoing accessions, of which 3469 and 6200 accessions, respectively, were for the genebank. Thus, the share of the SHU activities related to the genebank is estimated at 14% (9669 ÷ 69,915). Of the cost allocated to the genebank-related activities, 35% was allocated to testing incoming accessions and 65% was allocated to the dissemination of accessions.

Seed-health testing facility. The area of the SHU is 364 m² at a cost of $340/m² (15,000 pesos), making the total construction cost $123,760 (364 × $340). We allocated 14% of the facility cost to the genebank.

Laboratory and office equipment. Laboratory equipment includes two growing chambers at $9000 each, three refrigerators at $840 each, an incubator at $20,000, an autoclave at $15,000, 15 telescopes at $2500 each, a photoscope at $12,000, and other miscellaneous equipment at $10,000, amounting to $115,020 in total. Office equipment includes desks, chairs, copiers, cabinets and so on, totalling $10,000. Fourteen per cent of these costs was allocated to the genebank.

Computers. There are four computers with UPS devices costing $1620 each and two printers at $700 each, of which 14% is allocated to the genebank.

Vehicle. One pick-up truck costing $12,000 is assigned to the SHU, of which 14% is allocated to the genebank.

General capital

General facility. The remaining office and other laboratory area related to the genebank is 304 m^2 and costs $103,360 ($340 $\times$ 304).

Office equipment. Includes desks, tables, chairs, filing cabinets and so on, at a total cost of $30,000.

Computers. There are three servers with UPS devices costing $5700 each, 33 PCs with UPS devices at $1620 each, 11 printers at $700 each, and other computer-related equipment at $2000.

Vehicle. One vehicle is used for general genebank operation.

Table 6.3. Annual operating costs of conservation and distribution at the IRRI genebank

Representative labour costs (including salary and benefits) are as follows:

Internationally recruited scientist: $114,000
Senior scientist: $21,400
Scientist: $11,200
Secretary: $6700
Technician: $5000 (known as 'T-labour' at $20 per day)
Casual (daily) contractual worker: 190 pesos per day or $4.3 per day based on an exchange rate of 44 pesos to 1 US$ (known as 'C-labour')

Sixty per cent of the genebank head's time is spent on overall management of the genebank operation. The genebank manager and two crop scientists spend 90% of their time on genebank operations and regeneration. The overheads rate of 30.07% is adjusted to 27.14%.

Acquisition

Seed-health testing. The head of the entomology and plant pathology division oversees the operations of the SHU. Five per cent of the SHU head's time is spent managing the unit ($114,000 $\times$ 0.05); four scientists and eight technicians cost $90,500 ($114,000 $\times$ 0.05 + $11,200 $\times$ 4 + $5000 $\times$ 8); and office and laboratory supplies are estimated to cost $12,000. Fourteen per cent of these costs are allocated to the genebank, of which 35% represents testing incoming accessions and 65% represents disseminating accessions.

Seed handling. After the post-entry inspection in the SHU, incoming seeds are checked for duplicate samples, tested for viability (see below), sorted and stored in the cold-storage room. About 30 accessions can be processed per day by a technician, making the labour cost $3300 ($20 $\times$ 165 days). The genebank manager spends 10% of his/her time managing this process, representing $1926 ($19,260 $\times$ 0.1). A paper bag for each accession costs 3 cents, making the non-labour cost $149.

Medium- and long-term storage

Storage management. Twenty per cent of the genebank manager's time is spent on conservation management, costing $3852 ($19,260 × 0.2); this cost is allocated at a 3:2 ratio between medium- and long-term storage.

Climate control. One technician monitors and operates the storage rooms year-round. The electricity to operate the cooling system in the medium- and long-term storage rooms is 28 kWh and 21 kWh, respectively; the average operating time is 12 h/day, making the electricity consumption for medium-term storage 122,640 kWh (28 kWh × 12 × 365) and for long-term storage 91,980 kWh (21 kWh × 12 × 365). At 7.3 cents/kWh, the total electricity cost is $8953 for medium-term storage and $6715 for long-term storage.

Viability testing

Viability testing. One technician conducts the viability testing year-round with the help of two casual workers. Ninety per cent of their time is allocated to the genebank (the balance is used for INGER accessions). With the additional 10% of the genebank manager's time, the total labour cost is $8361 (0.9 × [$5000 + 2 × $1075] + $19,260 × 0.1). Laboratory costs include paper towels, chemicals and Petri dishes at $400. Ninety per cent of this material cost is allocated to the genebank.

Dissemination

Dissemination management. Twenty per cent of the genebank manager's time is allocated to managing dissemination of accessions at a ratio of 7:3 between dissemination and duplication.

Seed-health testing. Sixty-five per cent of the seed-health testing cost is allocated to the genebank. (See 'Acquisition'.)

Packing/shipping. Half of one technician's time is spent on germplasm dissemination at the ratio of 7:3 between dissemination and duplication. Two hundred accessions can be packed per 'T-day', for a total packing cost of $2480 ($20 × 124). Each accession is packed in a small (8 cm × 11 cm) aluminium bag costing 11 cents each, for a total cost of $682 ($0.11 × 6200). Most accessions are sent in manila envelopes, so with 186 shipments the cost of envelopes is $186 ($1 × 186). The cost of each phytosanitary certificate is 11 cents and, with 92 external shipments, the total certificate cost is $10. The shipping cost in 1999 was $2500.

Duplication

Packing/shipping. See 'Dissemination'. The cost of aluminium bags for duplication is $1040 ($0.11 × 9450), while the shipping cost to the NSSL in 1999 was $1064.

Information management

Database management. The data management of the genebank is undertaken by two scientists and one technician, costing $27,400 in total. Annual expenses for software licensing include Oracle at $1880, Control IT at $88 and McAfee at $240; the total non-labour cost is $2208. We allocated 80% of the information management cost here to exclude the costs of training, collection, characterization and evaluation operations.

Other expenses. This includes magnetic tapes and jazz and zip disks at a total cost of $3000, of which 80% is allocated to the genebank.

General management

Managerial staff. Sixty per cent of the genebank head's time is spent managing the genebank operation, representing $68,400 (0.6 × $114,000). Additional labour includes 70% of a secretary, costing $4690 (0.7 × $6700) and 50% of an office technician, costing $2500 (0.5 × $5000).

Electricity. An average of 100,000 kWh of electricity is consumed by the genebank operations, in addition to electricity for storage, making the total cost of electricity $7300 ($0.073 × 100,000). We allocated 70% of this cost to the genebank to exclude INGER and other research-related activities.

Other expenses. The estimated costs of supplies such as the telephone, stationery and so on is $20,000, of which 70% is allocated to the genebank.

Table 6.4. Annual costs of regeneration and characterization at the IRRI genebank

Regeneration

Field management. A field scientist manages cultivated rice, and a screenhouse scientist manages wild species and some cultivated rice in a screenhouse. Since 10% of their time is dedicated to INGER-related activities, we allocated 90% of these costs to conservation ($10,080 = $11,200 × 0.9). Sixty per cent of the field scientist's time is allocated to the regeneration of cultivated rice in the dry season and 40% to the field operation and characterization in the wet season. Sixty per cent of the screenhouse scientist's time is allocated to the regeneration of wild rice, 10% to characterization, 20% to the seed cleaning and 10% to the regeneration of cultivated rice, which requires special care in the screenhouse.

Seed preparation. Planting materials are first prepared and then cross-checked with planting lists and sorted for ready seeding. This step requires 10 T-days for cultivated rice and 2 T-days for wild rice. All incoming accessions are hot-water-treated to deal with a nematode, which requires 10 T-days for cultivated rice and 2 T-days for wild rice. One envelope costing 1 cent is used for each accession.

Embryo rescue. Labour for the embryo rescue directed to seed germination includes 5% of a scientist, costing $560, and 40% of a technician, costing $2000. The annual cost of chemicals and materials is $1500; the rental cost of using phytotron for germination purposes is $95 per year. Forty per cent of these costs are allocated to wild rice and 60% to cultivated rice.

Land preparation. The experiment station charges $390/ha for land preparation including ploughing, harrowing, levee construction and repair. Ten hectares are used for regeneration of cultivated rice, costing $3900. Seventy per cent of this cost is allocated as labour, leaving 30% as non-labour cost. Two types of seed-beds are used for seeding – dry beds in the dry season and wet beds in the wet season or for the initial seed multiplication. The experiment station charges $20/ha for dry-bed preparation, which can be done mechanically, and $260/ha for wet-bed preparation, which is prepared manually. In 1999, 4.5 dry-bed ha were prepared costing $90 ($20 × 4.5) and 5.5 wet-bed ha were prepared at $1430 ($260 × 5.5). We allocated 90% of these costs as labour and 10% as non-labour. In addition, five T-days were needed for the preparation of seed-beds for cultivated rice in the screenhouse. A seed-box rather than a seed-bed is used for the initial seeding of wild rice, and the preparation of seed-boxes requires 30 T-days.

Seeding. Direct seeding is used for *Oryza glaberrima* (250 accessions), which requires 10 T-days and 5 C-days, costing $221 (10 × $20 + 5 × $4.3). For *Oryza sativa* (7050 accessions), 20 T-days and 20 C-days are required, costing $486 (20 × $20 + 20 × $4.3). For wild rice, seeding requires 15 T-days.

Transplanting. The field needs to be laid out prior to transplanting, which requires 10 T-days and 30 C-days, costing $329 for cultivated rice (10 × $20 + 30 × $4.3), and 5 T-days for wild rice. Seedlings are then removed, requiring 20 T-days and 20 C-days at a cost of $486 for cultivated rice (20 × $20 + 20 × $4.3) and 10 T-days for wild rice. Finally, seedlings are transplanted and replanted, requiring 20 T-days. The experiment station provides casual workers to transplant and replant seedlings at a cost of $140/ha. Wild rice requires 20 T-days for replanting.

Pest control. Cultivated rice requires monitoring for pests, requiring 20 T-days; insect pest control, requiring 30 T-days and $498 in chemicals; roguing and disease control, requiring 20 T-days and $100 in chemicals; rat control, requiring 40 C-days and $2432 in materials; snail control, requiring 10 T-days, 30 C-days and $290 in chemicals; and bird watching (provided by the experiment station at a cost of $356/ha). Screenhouse cultivation requires monitoring (5 T-days), insect pest control (5 T-days and $50 in chemicals), roguing and disease control (5 T-days and $5 in chemicals), rat control (5 T-days) and snail control (5 T-days and $10 in chemicals). Ninety per cent of the cost of screenhouse cultivation is allocated to wild rice, and 10% to cultivated rice.

Irrigation/fertilization. Irrigation required 20 T-days for cultivated rice and 5 T-days for wild rice; the water cost is $86/ha. Spraying chemicals requires 20 T-days and 40 C-days for cultivated rice and 10 T-days for wild rice. The total cost of fertilizers for cultivated rice (ammonium sulphate, muriate of potash, solophos, zinc and so on) is $1570; the cost for wild rice is $150.

Purification. Cultivated rice requires 30 T-days and 20 C-days for purification; wild rice requires 10 T-days. Wild rice also needs bagging to prevent shattering, which requires 30 T-days. The cost of bags is 10 cents per accession (one bag).

Harvesting. For cultivated rice, cutting and hauling requires 150 T-days and 800 C-days, while panicle harvesting requires 30 T-days and 60 C-days. For wild rice, panicle harvesting requires 80 T-days. Harvested materials are put in cloth bags and are then transported to the seed-drying room. The cost of bags is 10 cents each.

Threshing/cleaning. Cultivated rice requires 150 T-days and 200 C-days for threshing and blowing, while wild rice requires 30 T-days.

Seed processing

Process management. Forty per cent of the genebank manager's time is allocated to managing seed processing, at a cost of $7704 (0.4 × $19,260). Based on the number of processed accessions, 94% of this cost was allocated to cultivated rice and 6% to wild rice.

Seed drying/cleaning. Prior to drying, 100 accessions can be transferred from cloth bags to paper bags (which cost 5 cents each) in one T-day. The electricity to operate the drying equipment is included in the general costs in Table 6.3. One T-day is needed to clean two accessions of cultivated rice or three accessions of wild rice. Given 7400 accessions of cultivated rice, 3700 T-days are needed, totalling $74,000 (3700 × $20); cleaning 166 accessions of wild rice costs $3320 (166 × $20). In addition, a screenhouse scientist spends 20% of her time cleaning wild rice, costing $2016 (0.2 × $10,080).

Medium-term/long-term packing. Two hundred medium-term or 100 long-term accessions can be packed in 1 T-day.

Characterization

The characterization of cultivated rice is undertaken in the wet season. Two thousand accessions were planted on 2 ha of land in the 1999 growing season. The characterization of wild rice is undertaken simultaneously with regeneration in the screenhouse.

Field operation. The field operation for characterization in the wet season is similar to that for regeneration but there is no harvesting.

Recording traits. Twenty per cent of the field scientist's time and 100 T-days are required to record cultivated rice traits. For wild rice, 10% of the screenhouse scientist's time and 30 T-days are required to record traits.

Chapter 7: CIAT Genebank

(All costs are in US dollars, 2000 prices. Costs were converted to US dollars at a rate of 2000 pesos per dollar.)

Table 7.2. Capital input costs for cassava at the CIAT genebank

Storage

Conservation facility. The size of the *in vitro* conservation room is 45 m^2 at a construction cost of $481/m^2, making the total cost $21,645.

Climate-control equipment. The conservation room houses two small dehumidifiers at $500 each and a central air-conditioner at $3000, of which 80% is allocated to the conservation room and 20% to the growing room located next door.

Cryoconservation tank. This category includes three conservation tanks at $5000 each, requiring six racks per tank at $87 each. The tank takes up very little space; hence we overlooked a building cost. Three test tubes at 18 cents each are required per accession, for a total cost of $1080 for the estimated 2000 accessions.

Other equipment. This includes six fixed shelves at $500 each and 540 plastic trays at $2 each installed in the conservation room to hold test tubes. Each *in vitro* accession is stored in five test tubes costing 22 cents each. With 6080 accessions, the total cost of test tubes is $6688. (Note that test tubes needed for cryoconservation are included with the cost of the cryoconservation tank, above.)

Subculturing

Subculturing facility. The facility for subculturing includes a growing room of 20 m^2, a subculturing room of 13 m^2, and a washing room of 12 m^2, for a total of $21,645.

Laboratory equipment. The total cost of laboratory equipment is $52,530, including microscopes, an agitator, a germination chamber, a sterilizer, incubators and miscellaneous furniture.

Other equipment. This category includes a third of the $3873 cost of a bar-code system ($11,620 × 0.33), furniture costing $700, shelves and trays in the growing room at $2040, and 20% of a central air-conditioner and dehumidifiers in the growing room at $800 ($4000 × 0.2). Five small glass tubes per accession, at 8 cents each, are used for subculturing in the growing room, representing a total cost of $1716 for 4290 subculturing accessions.

Disease indexation

Testing facility. The Germplasm Health Laboratory (GHL) is 136 m², of which one-third is used for cassava disease indexing.

Greenhouse. Of four greenhouses, with a construction cost of $26,000 each including equipment, 1.5 greenhouses are used for the cassava disease indexing.

Laboratory/office equipment. The total equipment cost for the Germplasm Health Unit (GHU) is $145,850; other office equipment adds $15,500, of which one-third was allocated to cassava disease indexing.

Computers. The GHU houses two computers costing $2500 each and a printer costing $1500, of which one-third is allocated to conservation activities through the genebank.

Cryoconservation

Cryoconservation laboratory. The presumed building area for the cryoconservation operation is 15 m², costing $7215.

Laboratory equipment. Most equipment for cryoconservation operation is shared with the *in vitro* conservation operation. We assumed that 20% of the cost of lab and other equipment for subculturing is required for cryoconservation.

General capital

General facility. An office of 20 m² and a machine room of 24 m², each costing $120/m², are used exclusively for *in vitro* conservation. In addition, 25% of the general office area (250 m²) is allocated to *in vitro* conservation. Another 5% of the general office area is allocated equally between cryoconservation and the field genebank.

Other equipment. This includes office equipment at $15,000 and a herbarium at $8000, of which 25% was allocated to *in vitro* conservation and 5% equally between cryoconservation and the field genebank.

Computers. The GRU uses 16 computers at $2500 each and seven printers at $1500 each, of which 25% was allocated to *in vitro* conservation and 5% equally between cryoconservation and the field genebank.

Table 7.3. Capital input costs for conserving bean and forage at the CIAT genebank

Medium-term storage

Storage facility. With a construction cost of $481/m^2, the cost of the medium-term storage room, at 120 m^2, is $57,720. Next to the storage and drying rooms is a common transition room of 11 m^2 that houses control equipment. The cost of the transition room is allocated equally among medium-term storage, long-term storage and the final drying rooms.

Storage equipment. Includes a cooling system ($15,700), a dehumidifier ($8800), insulation equipment ($25,000) and a third of the equipment in the transition room ($3130 = 0.33 × $9400).

Other equipment. This category includes 17 mobile shelving units at a cost of $25,000, 3642 metal trays at a cost of $14,370 and one-third of a bar-code system at a cost of $11,620.

Seed containers. One plastic jar (containing 5000–30,000 seeds) is used for most accessions, with the exception of a few beans with large-sized seed that require two jars. Currently, a total of 50,820 jars for beans, 18,180 for forages and 1000 for wild cassavas are housed in the medium-term storage room. With a unit cost $1.10 per jar, the total container cost is $77,000.

Long-term storage

Storage facility. The facility cost of the long-term storage room, at 86 m^2, is $41,366. One-third of the cost of a transition room, at 11 m^2, is also allocated to the storage facility.

Storage equipment. This includes a freezer at $21,000, insulation equipment at $20,000 and one-third of the equipment in a transition room at $3130 (0.33 × $9400).

Other equipment. This includes 12 mobile shelving units at a cost of $18,250, 2484 metal trays at a cost of $9720 total and one-third of a bar-code system at a cost of $11,620.

Seed containers. One to five aluminium bags (usually one or two) are used for long-term conservation. In addition, three to six additional bags are stored for viability, repatriation and safety duplication. With a unit cost of

20 cents per bag and assuming an average of six bags per accession (two for storage, one for viability, one for repatriation and two for safety duplication), the total container cost is $24,108 (based on 12,700 bean and 7390 forage accessions).

Viability testing

Viability-testing facility. Forage testing is undertaken in a 27 m² lab, and bean testing is done in a 120 m² working area. The total facility cost is $27,387, representing $12,987 for the lab and $14,400 for the working area.

Testing equipment. The total equipment cost inside the lab is $21,155, including germination chambers, scales, incubators and so on, and the sand-beds for bean testing in the working area cost $1390.

Regeneration

Greenhouse. Two GRU greenhouses costing $26,000 each are used for bean regeneration. (Other greenhouses/mesh-houses used by the genebank are charged on a lease-basis by CIAT headquarters, and are costed accordingly as operating costs.) Five mesh-houses costing $13,000 each and four 'coverts' costing $10,200 in total are used for bean regeneration in Popayan and were built by the GRU.

Field equipment. This includes four tractors costing $13,000 each, two cutting machines costing $1400 each, and a few bicycles costing $300.

Seed-drying facility. There are two drying facilities, with a total area of 34 m². A third of the cost of a transition room (11 m²) is also costed here.

Seed-drying equipment. There are three drying cabinets costing $10,000 each. Equipment in the drying room costs $24,000, including the cooling system, dehumidifier, shelving and so on. The final drying room costs $51,200, and one-third of the equipment cost in a transition room is also included.

Seed-processing facility. This includes a seed processing/cleaning/packing area of 121 m² and a temporary storage room of 55 m².

Seed-processing equipment. This includes external seed-threshing equipment, costing $13,520; internal cleaning and processing equipment, including scales, a seed blower, a thresher and so on, costing $49,900 in total; and a cooling system and other equipment in the temporary storage room, costing $34,520.

Seed-health testing

Seed-health testing facility. The total area of the GHU is 136 m², which we allocated equally among cassava, beans and forages.

Greenhouse. The cost of a greenhouse and cooling system is $26,000, of which half is used for post-quarantine planting.

Laboratory equipment. The total equipment cost for the GHU is about $145,850; other office equipment costs $15,500, of which two-thirds was allocated to bean and forage health testing.

Computers. The GHU uses two computers at $2500 each and a printer at $1500, of which two-thirds was allocated in this cost category.

General capital

General facility. Unallocated space in the genebank facility includes the office, herbarium and other space at 250 m^2, of which 70% was allocated to bean and forage conservation.

Other equipment. Office equipment includes desks, tables, telephones and other office equipment, costing $15,000; the equipment for the herbarium costs $8000, of which 70% is allocated here.

Computers. The GRU uses 16 computers at $2500 each, one server at $5000, and seven printers at $1500 each, of which 70% is allocated here, with the balance allocated to cassava conservation.

Table 7.4. Annual operating costs for cassava at the CIAT genebank

Representative annual labour costs, including benefits, at CIAT are as follows:
　　Genebank head (PhD level): $125,400
　　Associate (MA level): $28,800
　　Assistant (BA level): $13,700
　　Secretary (BA level): $14,700
　　Technician: $8600
　　Worker: $7400
　　Temporary worker: $21 per day

The overheads rate at CIAT is 22.1%.

Storage
(Note that field genebank operation is explained under 'Field maintenance'.)

Storage management. Ten per cent of an associate's labour is allocated to regular monitoring and managing *in vitro* accessions, costing $2880 ($28,800 × 0.1). For cryoconservation, 5% of an assistant scientist's time is allocated to regular monitoring.

Electricity/supplies. Electricity to run equipment for *in vitro* conservation is $5443 (7 cents × 77,760 kWh). Supplies for various operations used in *in vitro* conservation cost $2539. We allocated 40% of this cost to storage, 50%

to subculturing, and 10% to dissemination. For cryoconservation, 3 t of liquid nitrogen costing $300 is required per year.

Subculturing

Management. Forty per cent of an associate's time is allocated to monitoring and managing the subculturing activity for *in vitro* conservation, costing $11,520 ($28,800 × 0.4).

Subculturing. One technician and 2.8 workers are required for subculturing for conservation and for dissemination. We allocated 90% of the labour to subculturing, costing $26,388 (0.9 × [$8600 + 2.8 × $7400]), and the remaining 10% to dissemination. Fifty per cent of the costs of chemicals and supplies were allocated to subculturing, costing $1270 ($2539 × 0.5). The cost of electricity to run the subculturing equipment is $4727, of which 90% was allocated to subculturing (and the balance to dissemination).

Viability testing. The viability testing of 300 accessions for cryoconservation requires 2 months of a technician's time and laboratory materials costing $100.

Disease indexation

Management. Five per cent of an associate's time in the GHU is allocated to the management of cassava disease indexing. One-third of a half-time secretary based in the quarantine unit is also allocated to this category.

Greenhouse operation. A full-time worker manages cassava plants grown in the greenhouse. The cost of chemicals used in the greenhouse is $3532.

Laboratory operation. Ninety per cent of an assistant scientist's time and 50% of a worker are allocated to various types of disease indexing. The costs of chemicals/supplies for grafting, ELISA and PCR are $350, $6680 and $15,070, respectively.

Field maintenance

Field management. Five per cent of an IRS scientist and a full-time secretary in the cassava breeding programme along with the programme's annual operating expenses ($15,000) was allocated to field management.

Land preparation. Land preparation is undertaken by the field operation unit, which charges $650 for a 5 ha area, of which 70% is allocated as labour and 30% as non-labour. The labour cost is $2115 for the preparation of stakes for planting, and chemicals and materials cost $286.

Planting. The labour and non-labour costs of planting are $2115 and $352, respectively.

Field maintenance. The labour cost for weed control includes $895 for spraying and $1415 for hand-weeding, and the cost of chemicals is $806. For pest control, the labour cost is $424; the non-labour cost (for chemicals) is $1185; and the cost for biological materials is $32. For disease control, the labour cost is $118 and the chemical cost is $167. The labour cost of pruning and logging is $420. In addition, the irrigation cost at $473 is charged by the field operation unit. We allocated 90% of this to labour.

Harvesting. The labour cost for harvesting is $3410; the material cost is $740.

Characterization. Labour to record traits and enter data costs $3182.

Cryo-operation

Management. Thirty per cent of an assistant's time is allocated to the operation of cryoconservation. (This process is currently performed at the biotechnology unit, using its resources and equipment; however, when the operation is moved to the *in vitro* conservation unit in the GRU facility, the additional cost of equipment and materials is expected to be low. In addition to a small preparation space, about 20% of the cost of equipment and materials for *in vitro* conservation will be required for cryoconservation.)

Processing. Two technicians can process 1000 accessions/year. The cost of chemicals and other supplies is estimated to be about 20% of the cost for subculturing ($1105).

Dissemination

Management. Ten per cent of an associate's time is allocated to managing dissemination.

Subculturing. To adjust for exceptional demand in 2000, we allocated 30% (rather than 20%) of the labour and non-labour costs of subculturing.

Packing/shipping. Twenty per cent of a worker's time is required for packing. Five glass tubes costing 8 cents each are disseminated per external accession; the total tube cost is $96 for 240 accessions. Of 54 shipments, 20 were shipped externally from CIAT. The average shipping cost was $30, making the total shipping cost $600. The cost of shipping boxes is $1 per shipment.

General management

Managerial staff. Ten per cent of an associate's time is allocated to data management for the *in vitro* collection. Fifteen per cent of the genebank head's time and a full-time secretary is allocated to the *in vitro* collection; 5% is allocated to cryoconservation and the field genebank.

Office expenses. Office expenses for *in vitro* conservation cost $2524. Twenty-five per cent of general office and electricity expenses, costing $29,546, are also allocated to *in vitro* conservation, and 5% are allocated to cryoconservation and the field genebank.

Table 7.5. Annual operating costs for bean and forage at the CIAT genebank

Medium- and long-term storage

Storage management. An associate manages the conservation group in the GRU. Fifteen per cent of her time is spent on storage management (10% for medium-term and 5% for long-term). The allocation between beans and forages is based on the number of accessions stored.

Temperature control. Four months of a technician's time from the engineering unit is allocated to monitoring equipment in both the medium- and long-term storage rooms. The annual cost of electricity to operate the cooling system is $3050 for medium-term and $2325 for long-term storage. These figures are allocated between beans and forages based on the number of accessions stored.

Viability testing

Viability testing. Five per cent of an associate's time is allocated between beans and forages, based on the number of accessions tested. A full-time technician tests bean accessions, and 70% of a worker's time is required to test forage accessions. Bean testing requires treated sand at a cost of $130 per year; forage testing requires chemicals, at $696, other materials, at $226, and electricity to operate the equipment, at $1000.

Regeneration

Field operation. For beans, an assistant manages the field activities undertaken by three technicians, three workers and two temporary workers. In addition, during planting and harvesting, additional temporary labour costing $780 was used. Non-labour costs include land rental and preparation in Tenerife, costing $700 for 3 ha; chemicals costing $2656; other materials costing $1625; one vehicle costing $2800; travel costing $2380; and rent for four mesh-houses in Palmira costing $11,200 (4 × $2800). For forages, a technician manages the field activities undertaken by one technician, one worker and six temporary workers. Non-labour costs include land cleaning at $1280 ($20/ha × 8 ha × 8 times per year), machine maintenance at $700, chemicals at $2545, additional materials at $1500, one vehicle at $2800, travel at $1440 and rental for four mesh-houses in Palmira at $11,200.

Seed processing. An associate spends 20% of his/her time managing seed processing (allocated based on the number of accessions). For beans, two temporary workers are needed for threshing/cleaning, costing $10,080 (2 × $5040) and two regular workers are needed for processing and packing at a cost of $14,800 (2 × $7400). For forages, one technician and one temporary worker are needed for cleaning, costing $13,640 ($8600 + $5040) and 60% of a technician is needed for processing and packing at a cost of $5160 (0.6 × $8600). Non-labour costs include electricity for drying and processing, along with temporary storage, at a cost of $15,755 ($10,582 + $2330 + $2843) and other expenses of $1000, all of which are allocated based on the number of accessions.

Dissemination

Dissemination management. Twenty per cent of an associate's time is allocated to the management of dissemination, allocated between beans and forages based on the number of accessions disseminated.

Seed-health testing. The labour in the GHU for beans and forages includes 60% of the unit head's time, 90% of an assistant's time, 33% of a secretary, and 150% of a worker, resulting in a quasi-fixed cost of $17,280 and a labour cost of $59,825. The annual cost of chemicals and other supplies is $12,000. Ninety per cent of these costs were allocated to genebank accessions (and the balance to the breeding programme). About 5000 bean and 1500 forage accessions are tested annually. Thus, the per-accession costs of quasi-fixed inputs, labour and non-labour are $2.40, $8.28 and $1.66, respectively. We multiplied these costs by the number of disseminated accessions.

Packing/shipping. A worker can pack 100 accessions per day at a cost of $893 for beans ($21 × 42.5) and $109 for forages ($21 × 5.2). The cost of a small aluminium bag is 10 cents, and the cost of a box for shipping is $1. For beans, the cost of bags is $425 and boxes cost $25, based on 25 external shipments. For forages, the corresponding figures are $52 and $17, respectively. The shipping costs are $180 for beans and $80 for forages.

Duplication
In 2000, a total of 2180 bean accessions were prepared for safety duplication to Brazil and Costa Rica. Duplication only occurs once every few years so it is difficult to derive an exact number of accessions and corresponding costs. Hence the following figures are based on standard operations and costs.

Packing/shipping. A worker can pack 100 accessions per day so 21.8 days are required to pack 2180 accessions representing a total cost of $458 ($21 × 21.8). The total cost of aluminium bags is $218 (2180 × 10 cents) and the shipping cost is $200 for two shipments.

General management

Managerial staff. Forty per cent of the genebank head's time and a full-time secretary's time are equally allocated between beans and forages. Thirty per cent of an associate's time and 40% of a technician's time are allocated equally between beans and forages for data management.

Office expenses. Office expenses include 12 computers at $2640 ($220 × 12), 16 telephones costing $5184 ($324 × 16), five printers costing $1250 ($250 × 5), and one vehicle costing $2800. Other miscellaneous expenses amount to $15,000, and electricity costs $2672. We allocated 70% of these costs equally to beans and forages, leaving 30% allocated to cassava.

References

Alston, J.M., Marra, M.C., Pardey, P.G. and Wyatt, T.J. (2000) *A Meta Analysis of Rates of Return to Agricultural R&D: Ex Pede Herculem?* IFPRI Research Report No. 113, International Food Policy Research Institute, Washington, DC.

Anderson, J.R. and Dalrymple, D.G. (1999) *The World Bank, the Grant Program, and the CGIAR: A Retrospective Review.* OED Working Paper Series No. 1, March, Operations Evaluation Department, World Bank, Washington, DC.

Appa Rao, S., Bounphanouxay, C., Phetpaseut, V., Schiller, J.M., Phannourath, V. and Jackson, M.T. (1997) Collection and preservation of rice germplasm from southern and central regions of the Lao PDR. *Lao Journal of Agriculture and Forestry* 1, 43–56.

Baum, W.C. (1986) *Partners Against Hunger: the Consultative Group for International Agricultural Research.* World Bank, Washington, DC.

Beattie, B.R. and Biggerstaff, D.R. (1999) Karnal bunt: a wimp of a disease but an irresistible political opportunity. *Choices* second quarter, 4–8.

Berthaud, J., Savidan, Y., Barré, M. and Leblanc, O. (1997) Tripsacum. In: Fuccillo, D., Sears, L. and Stapleton, P. (eds) *Biodiversity in Trust: Conservation and Use of Plant Genetic Resources in CGIAR Centers.* Cambridge University Press, Cambridge, UK, pp. 227–233.

Burstin, J., Lefort, M., Mitteau, M., Sontot, A. and Guiard, J. (1997) Towards the assessment of the cost of genebanks management: conservation, regeneration, and characterization. *Plant Varieties and Seeds* 10, 163–172.

El-Ahmed, A. (1998) International Center for Agricultural Research in the Dry Areas. In: Kahn, R.P. and Mathur, S.B. (eds) *Containment Facilities and Safeguards for Exotic Plant Pathogens and Pests.* American Phytopathological Society, St Paul, Minnesota, pp. 45–53.

Ellis, R.H. and Jackson, M.T. (1995) Accession regeneration in genebanks: seed production environment and the potential longevity of seed accessions. *Plant Genetic Resources Newsletter* 102, 26–28.

Epperson, J.E., Pachico, D. and Guevara, C.L. (1997) A cost analysis of maintaining cassava plant genetic resources. *Crop Science* 37, 1641–1649.

Escobar, R.H., Mafla, G. and Roca, W.M. (1997) A methodology for recovering cassava plants from shoot tips maintained in liquid nitrogen. *Plant Cell Reports* 16, 474–478.

FAO (Food and Agriculture Organization of the United Nations) (1996) *Global Plan of Action for the Conservation and Sustainable Utilization of Plant Genetic Resources for Food and Agriculture.* FOA, Rome.

FAO (Food and Agriculture Organization of the United Nations) (1998) *The State of the World's Plant Genetic Resources for Food and Agriculture.* FAO, Rome.

FAO/IPGRI (Food and Agriculture Organization of the United Nations and International Plant Genetic Resources Institute) (1994) *Genebank Standards.* FAO/IPGRI, Rome.

Fowler, C. (2003) The status of public and proprietary germplasm and information: an assessment of recent developments at FAO. *IP Strategy Today* No. 7–2003. Available at: www.bioDevelopments.org

Frankel, O.H., Brown, H.D. and Burdon, J.J. (eds) (1995) *The Conservation of Plant Biodiversity.* Cambridge University Press, Cambridge, UK.

Fuentes-Davila, G. (1996) Karnal bunt. In: Wilcoxson, R.D. and Saari, E.E. (eds) *Bunt and Smut Diseases of Wheat: Concepts and Methods of Disease Management.* CIMMYT, Mexico, DF, pp. 26–32.

Furman, B.J., Qualset, C.O., Skovmand, B., Heaton, J.H., Corke, H. and Wesenberg, D.M. (1997) Characterization and analysis of North American triticale genetic resources. *Crop Science* 37 (6), 1951–1959.

Gollin, D., Smale, M. and Skovmand, B. (2000) Searching an *ex situ* collection of genetic resources. *American Journal of Agricultural Economics* 82, 812–827.

Gryseels, G. and Anderson, J.R. (1991) International agricultural research. In: Pardey, P.G., Roseboom, J. and Anderson, J.R. (eds) *Agricultural Research Policy: International Quantitative Perspectives.* Cambridge University Press, Cambridge, pp. 309–339.

Harlan, J.R. (1972) Genetics of disaster. *Journal of Environmental Quality* 1 (3), 212–215.

Hayami, Y. and Ruttan, V.W. (1985) *Agricultural Development: an International Perspective,* revised and expanded edition. Johns Hopkins University Press, Baltimore, Maryland.

Jackson, M.T. (1997) Conservation of rice genetic resources: the role of the international rice genebank at IRRI. *Plant Molecular Biology* 35, 61–67.

Jackson, M.T. (2000) *Genetic Resources Center: Manual of Operations and Procedures of the International Rice Genebank.* International Rice Research Institute, Los Baños, Philippines.

Jackson, M.T., Javier, E.L. and McLaren, C.G. (2000) Rice genetic resources for food security: four decades of sharing and use. In: Padolina, W.G. (ed.) *Plant Variety Protection for Rice in Developing Countries.* Limited proceedings of the workshop, 'The Impact of Sui Generis Approaches to Plant Variety Protection in Developing Countries', held 16–18 February 2000, at the International Rice Research Institute, Los Baños, Philippines. IRRI, Makati City, Philippines, pp. 3–8.

Kameswara Rao, N. and Bramel, P.J. (eds) (2000) *Manual of Genebank Operations and Procedures.* Technical Manual No. 6, International Crops Research Institute for the Semi-Arid Tropics, Patancheru, India.

Kameswara Rao, N. and Jackson, M.T. (1996) Seed longevity of rice cultivars and strategies for their conservation in genebanks. *Annals of Botany* 77, 251–260.

Koo, B. and Wright, B.D. (2000) The optimal timing of evaluation of genebank accessions and the effects of biotechnology. *American Journal of Agricultural Economics* 82 (4), 797–811.

Koo, B., Pardey, P.G., Qian, K. and Zhang, Y. (2003a) *The Economics of Generating and Maintaining Plant Variety Rights in China.* EPTD Discussion Paper No. 100, February, International Food Policy Research Institute, Washington, DC.

Koo, B., Pardey, P.G. and Wright, B. (2003b) The price of conserving agricultural bio-diversity: commentary. *Nature Biotechnology* February, 126–128.

Naredo, M.E.B., Juliano, A.B., Lu, B.R., de Guzman, F. and Jackson, M.T. (1998) Responses to seed dormancy-breaking treatments in rice species (*Oryza* L.). *Seed Science and Technology* 26, 675–689.

NRC (National Research Council) (1972) *Genetic Vulnerability of Major Crops.* Committee on Genetic Vulnerability of Major Crops. National Academy of Sciences, Washington, DC.

OECD (Organization of Economic Cooperation and Development) (2000) *Main Economic Indicators.* CD-ROM, OECD, Paris.

Pardey, P.G., Alston, J.M., Christian, J.E. and Fan, S. (1996) *Hidden Harvest: US Benefits from International Research Aid.* IFPRI Food Policy Report, September, International Food Policy Research Institute, Washington, DC.

Pardey, P.G., Koo, B., Wright, B.D., van Dusen, M.E., Skovmand, B. and Taba, S. (1999) *Costing the* Ex Situ *Conservation of Genetic Resources: Maize and Wheat at CIMMYT.* EPTD Discussion Paper 52, IFPRI, Washington, DC.

Pardey, P.G., Koo, B., Wright, B.D., van Dusen, M.E., Skovmand, B. and Taba, S. (2001) Costing the conservation of genetic resources: CIMMYT's *ex situ* maize and wheat collection. *Crop Science* 41 (4), 1286–1299.

Pardey, P.G., Alston, J.M., Chan-Kang, C., Magalhães, E.C. and Vosti, S.A. (2002) *Assessing and Attributing the Benefits from Varietal Improvement Research: Evidence from Embrapa, Brazil.* EPTD Discussion Paper No. 95, August, International Food Policy Research Institute, Washington, DC.

Pistorius, R. (1997) *Scientists, Plants and Politics: A History of the Plant Genetic Resources Movement.* International Plant Genetic Resources Institute, Rome.

Reid, R. and Konopka, J. (1988) IBPGR's role in collecting maize germplasm. In: *Recent Advances in the Conservation and Utilization of Genetic Resources. Proceedings of the CIMMYT Global Maize Germplasm Worksho*p. CIMMYT, Mexico, DF, pp. 9–16.

Reznik, S. and Vavilov, Y. (1997) The Russian scientist Nicolay Vavilov. Preface to the English translation of *Five Continents* by N.I. Vavilov. International Plant Genetic Resources Institute, Rome.

Roca, W.M., Rodrigues, J.A., Maflan, G. and Roa, J. (1984) *Procedures for Recovering Cassava Clones Distributed* In Vitro. CIAT Genetic Resource Unit, Cali, Colombia.

Sackville Hamilton, N.R. and Chorlton, K.K. (1997) *Regeneration of Accessions in Seed Collections: a Decision Guide.* Handbook for Genebanks No. 5, International Plant Genetic Resources Institute, Rome.

Salhuana, W.L., Pollak, M., Ferrer, M., Paratori, O. and Vivo, G. (1998) Breeding potential of maize accessions from Argentina, Chile, USA and Uruguay. *Crop Science* 38, 866–872.

SGRP (CGIAR System-wide Genetic Resources Programme) (1996) *Report of the Internationally Commissioned External Review of the CGIAR Genebank Operations.* International Plant Genetic Resources Institute, Rome.

SGRP (CGIAR System-wide Genetic Resources Programme) (2000) *A Funding Plan to Upgrade CGIAR Centre Genebanks.* Report of CGIAR System-wide Genetic Resources Programme, March, International Plant Genetic Resources Institute, Rome.

Smale, M. and Day-Rubenstein, K. (2002) The demand for crop genetic resources: international use of the US national plant germplasm system. *World Development* 30 (9), 1639–1655.

Smale, M., Reynolds, M.P., Warburton, M., Skovmand, B., Trethowan, R., Singh, R.P., Ortiz-Monasterio, I., Crossa, J., Khairallah, M. and Almanza, M. (2002) Dimensions of diversity in modern spring bread wheat in developing countries from 1965. *Crop Science* 42, 1766–1799.

Taba, S. (1997) Teosinte. In: Fuccillo, D., Sears, L. and Stapleton, P. (eds) *Biodiversity in Trust: Conservation and Use of Plant Genetic Resources in CGIAR Centers.* Cambridge University Press, Cambridge, UK, pp. 234–242.

Taba, S. and Eberhart, S. (1997) Cooperation between US and CIMMYT leads to rescue of thousands of Latin American maize landraces. *Diversity* 3 (1), 9–11.

Tanksley, S.D. and McCouch, S.R. (1997) Seed banks and molecular maps: unlocking genetic potential from the wild. *Science* 277, 1063–1066.

Valkoun, J., Robertson, L.D. and Konopka, J. (1995) Genetic resources at the heart of ICARDA mission throughout the Mediterranean region. *Diversity* 11 (1–2), 23–26.

Virchow, D. (1999) *Spending on Conservation of Plant Genetic Resources for Food and Agriculture: How Much and How Efficient?* ZEF Discussion Papers on Development Policy No. 16, September, Center for Development Research, Bonn.

Virchow, D. (2003) Current expenditures on crop genetic resources activities. In: Virchow, D. (ed.) *Efficient Conservation of Crop Genetic Diversity.* Springer-Verlag, Berlin, pp. 71–75.

Witt, S.C. (1985) *Biotechnology and Genetic Diversity.* California Agricultural Lands Project, San Francisco, California.

World Bank (2000) *World Development Indicators.* CD-ROM, World Bank, Washington, DC.

Wright, B.D. (1997) Crop genetic resource policy: the role of *ex situ* genebanks. *Australian Journal of Agricultural and Resource Economics* 41 (1), 81–115.

Zohrabian, A.G., Traxler, G., Caudill, S. and Smale, M. (2003) Valuing pre-commercial genetic resources: a maximum entropy approach. *American Journal of Agricultural Economics* 85 (2), 429–436.

Index

Page numbers in **bold** refer to figures; those in *italics* refer to tables or boxes. The letter 'n' is used where page numbers refer to the notes.

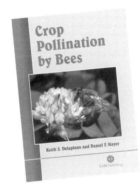
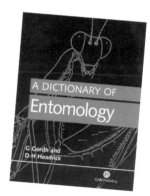
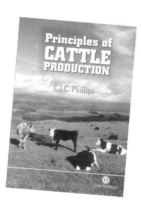